Graphic Problem Solving

for Architects & Builders

Graphic Problem Solving

for Architects & Builders

Paul Laseau

Acting Director, Ohio University School of Architecture

Cahners Books International, Inc.
221 Columbus Avenue
Boston, Massachusetts 02116

Library of Congress Cataloging in Publication Data

Laseau, Paul, 1937 –
   Graphic problem solving for architects and builders.

   1. Architectural drawing.   2. Architectural design.
I.   Title.
NA 2705.L37   720'.28   75-8607
ISBN 0-8436-0154-X

Second printing

ISBN: 0-8436-0154-X

Printed in the United States of America

The way we think affects our ability to solve problems. As an architect and an educator I have found that our thinking is dramatically altered when it is directly linked with drawings. Visual images have the power to alter our perceptions and on that simple fact hangs the rationale for graphic problem-solving and the inspiration for this collection of visual tools.

The long hours of mulling, searching and questioning that went into this book are dedicated to my parents who opened my eyes and my mind, and to my wife Peggy and children Michele, Kevin, and Madeleine, who bring me that joy of life that sustains me through each challenge.

I especially want to thank Forrest Wilson for his generous advice and encouragement, and James Monsul, Peter Hourihan, Geoffrey Egginton, Raymond Sluzas, Sheng P. Sheng and Arden Tewksbury who contributed many helpful suggestions.

Contents

Preface

Architects are aware of the importance of drawings as models of the building design problem. Drawings convey proposed buildings and even more important, assist the designer in the form of sketches, in developing and seeking solutions. In recent years architects and builders have become more concerned with the growing complexity of creating buildings and how the activities of a range of participants affect the final product. They are also concerned with how the building performs over time on many levels. These concerns present the architect, engineer or builder with a set of problems which are process- as well as product-oriented. I feel that the difficulties of these problems are often compounded by the laborious, pseudo-scientific way we are asked to approach them. Detailed user-need surveys, exhaustive traffic analysis, projected space demand can all be useful but often do not help us to "see" the essentials of a problem.

The concept of graphic problem-solving is based upon the recognition of an essential parallel between traditional building design problems and the new, emerging problems of building processes and building performance. This parallel is that in both cases there is a basic need for us to "see" problems in their simplest abstract form in order to best apply our mental abilities. In forming an abstract picture of a problem, graphics have the unique attribute of promoting a dialogue between the image and ourselves. We are not forced to consider the diagrams literally for they are open to several interpretations and manipulations. The graphic image is also efficient in that it makes possible the consideration of many variables and their interrelation at the same time. Finally the graphic image is often easily retained by our memory.

In this book I have illustrated four basic types of graphic or visual devices: the bubble diagram, area diagram, matrix and network. While not an exhaustive set of possible diagrams, these offer a degree of flexibility in their use, and the examples deal with the most common issues in visual problems: size, location, identity, relationship and process.

The reader is probably familiar with some of these devices in their more complex forms and may even be put off by the complicated uses to which they have often been put. It is my hope that the illustrations in this book will reveal the simplicity, efficiency and accessibility of visual tools for the problems we face each day. If this book can encourage the reader to consider graphics as a part of problem solving, it will have been successful.

As Peter Drucker put it in his book <u>Concept of the Corporation</u>, "This is true of all human organizations. Being human they can never aspire to perfection and must then make imperfection workable."

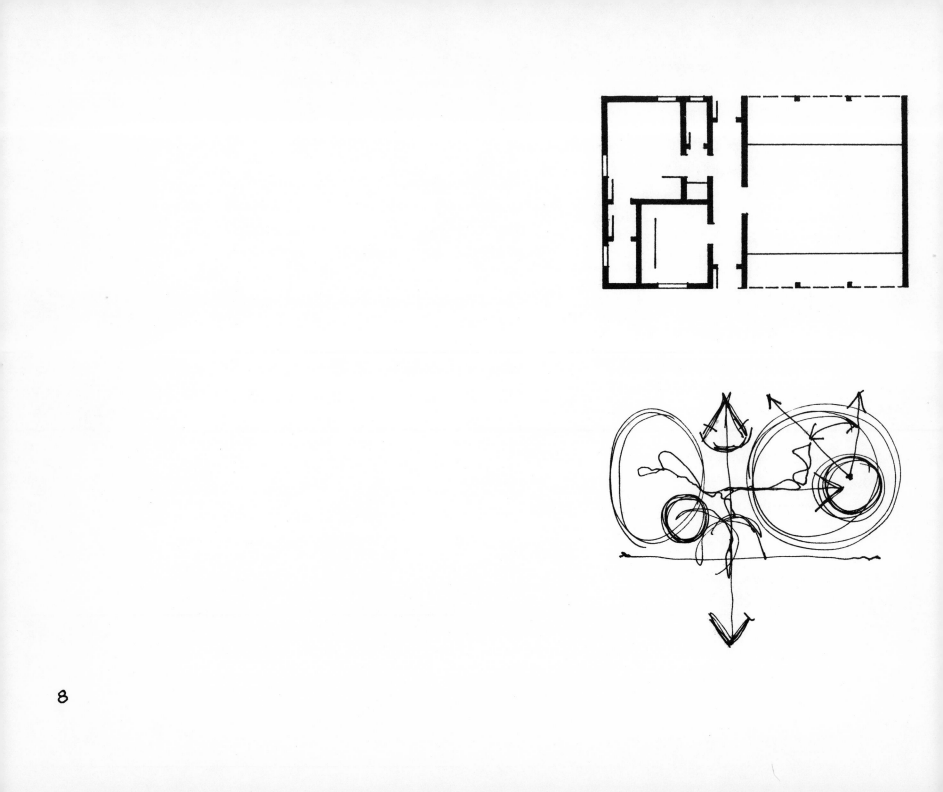

8

Introduction

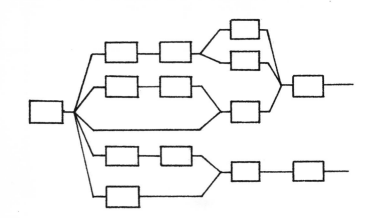

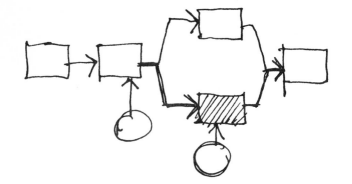

The general public thinks of architects as the the people who make drafted plans of buildings; in other words, designing and drawing plans are the same. This conception is at once amusing and understandable to architects. The real designing or thinking, of course, takes place amid a sea of freehand sketches which are remnants of a critical, visual conversation with ourselves. Not only has the design sketch been important throughout the history of architecture, at times it has been the only form of drawing used. The drafted plan has its uses but we are also aware of its limitations. With the growing complexity of programs and building processes, new approaches to building problems have tended to be expressed in very formal terms. How is it that we, as architects and builders, cannot see beyond these drawings to the thinking they represent? Why do we have a limited view as well? Understanding the reasons should help to clarify the objectives of this book.

9

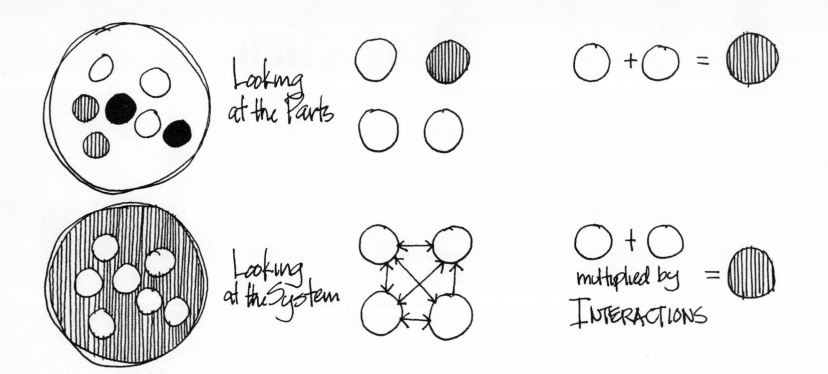

Looking at the Parts

Looking at the System

multiplied by INTERACTIONS

SYSTEMS THEORY AND VISUAL COMMUNICATION

A major obstacle to our adoption of process thinking is the mass of jargon and iconography used by many who promote the process approach. In trying to establish a new identity they lose effective contribution to architecture. The result is a misunderstanding within the building industry of the basic force behind the emerging problem-solving methods. Systems theory is a philosophy that holds simply that the essence of any organization or system lies as much or more in the interaction between its parts as in the nature of each part. It is important to note that architecture has the longest tradition of systems thinking. "The whole is greater than the sum of the parts" has always been a basic concept of architectural design, and naturally so, for we have always had to deal with form _and_ function, big _and_, small, technology _and_ people. What is more important is that as systems thinkers architects have developed a unique form of thinking —

The Building

Materials
Structure
Codes
Mechanical
Equipment
Costs
Program
Construction
People
Image

Medical
Services
Law
Recreation
Education
Industry
Communication
Energy
Housing

The Environmental
System

visual thinking. Through sketches the architect communicates with others and with himself in a special way that constantly focuses on the system characteristics of each problem. B. L. Whorf has theorized that language directly affects thinking. I am convinced that verbal language focuses on linear logic and detail while visual language is suited to simultaneous relationships and synthesis. Many people limit their thinking to the verbal, sequential level even when its effectiveness is becoming increa-

singly strained. We are "experiencing an information explosion [whose] volume threatens to diminish our capacity to absorb the essential information necessary to conduct our work in a purposeful manner," wrote Matthew Murgio. Our whole society, and in particular architects and builders, must understand the nature of relationships between parts and the whole. We need to see patterns in this complex world that will enable us to deal with its problems through effective application of our skills.

11

VISUAL THINKING

Because vision so dominates our learning experiences, it has always been a major factor in communication and in thinking. Man used graphic signs and symbols long before a written language was adopted. Symbols were a way of freezing ideas and events outside himself and of creating a history. Early written language itself was a highly specialized set of symbols derived from pictures that recorded events. Later the study of geometry developed diagrams that made possible thinking about structure, leading to the concept of design, the pre-modeling of objects constructed on a monumental scale. The creation of maps by piecing together the notes of explorers made it possible to make estimates of the unknown and stimulate new discoveries about both our world and the universe. The printing press with its ability to reproduce written language in quantity and with precision made possible specialization and the scientific revolution. In spite of the ascendency of the written word, visual perception has continued to be an essential part

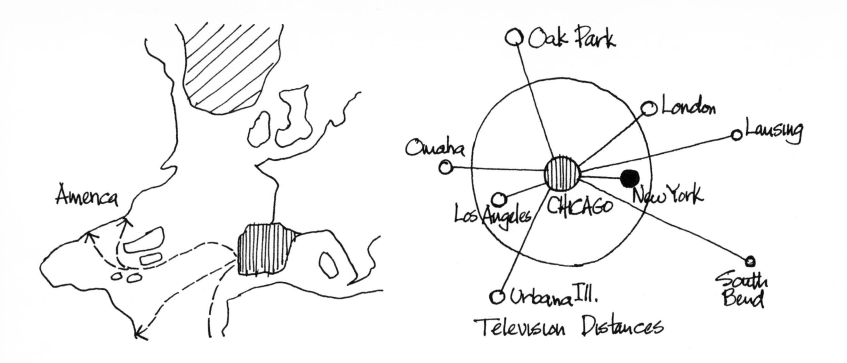

Television Distances

of our way of thinking. This is betrayed by the phrases: "I see what you mean; let's take a broader view; put the problem in perspective." And now visual thinking re-emerges in importance because we have come full circle. In the words of Marshall McLuhan, "After three thousand years of specialist explosion and increasing specialism and alienation in the technological extension of our bodies, our world has become compressional by dramatic reversal. As electrically contracted the globe is no more than a village."

A WARNING   This book contains plenty of material for producing "eye-wash." I have a strong prejudice on this point: Visual tools or graphics are _not a substitute for thinking._ We should have our fill by now of presentations dressed in "process graphics" that have plenty of arrows and no content. Along with the power to focus on concepts, graphics have the capability of distraction, to entertain as well as inform. It is time to quit treating visual tools as icons. Visual thinking is first of all _thinking._

13

The methods and tools for thinking found here, then, are not a panacea or magical truth but an extension of the architect's approach to new levels of concern. We look beyond our traditional problems in three ways:

1. Scale: an understanding of planning and policy issues of the larger physical context;
2. Program: investigation and application of relevant sociological and psychological research;
3. Process: consideration of the operations by which problems are solved surrounding and within the building industry.

It is important that we bring to bear on these issues the same level of visual (systems) thinking tools as were developed for traditional concerns. It is particularly necessary that these tools be developed for everyday conceptual, problem-solving thinking, and this is the emphasis of this book. Seeing a problem in its basic form before we are embroiled in its details is fundamental to effectiveness at every stage of the designing-building process because the cost of corrections or changes escalates as it progresses.

The following chapters of this book are collections of examples of the application of visual tools to problem-solving throughout the designing-building process. These examples are divided into simple categories starting with the most familiar and going to the least familiar. The concentration of examples around design functions is a result of both my background and the present state of visual thinking.

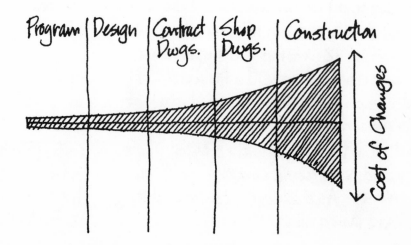

SCHEDULE OF EXAMPLES BY OPERATIONS IN DESIGN-BUILD PROCESS

	BUBBLE DIAGRAMS	AREA DIAGRAMS	MATRIX	NETWORKS
PLANNING	22,43	75,78,97	105	
DEVELOPMENT		78,88,97		138,139,151
MARKET ANALYSIS		76,88,97		
FEASIBILITY		64,86,94	118	
SITE - CONTEXT	39	62,67,68,76,84	118	
PROGRAM ANALYSIS	37,44,46,48	61,69,70,72,75,89	102	126,136,140
DESIGN MANAGEMENT	34,38,49	90,91	116	124,139,144
SCHEMATICS	28,30,35,38,40,42,50	65,80,82	108	126
PRELIMINARY DESIGN			101,106	152
PRESENTATION		60		128,130,150
DESIGN EVALUATION		73,91	113,114,119	
DESIGN DEVELOPMENT	52,53,54	74,96	103,104,110	144
PRODUCT SELECTION			101,115	151
CONSTRUCTION MANAGEMENT	41,57	74,77,93,94	100	134,141,142,143,148
CONSTRUCTION OPERATIONS	56			129,141
ENGINEERING		92,95,96		

15

Problem Solving

"No longer can we consider the artistic process as self contained, mysteriously inspired from above, unrelated and unrelatable to what people do otherwise." In the introduction to his book <u>Art and Visual Perception</u>, Rudolph Arnheim expressed the objective for this chapter. There is indeed a process by which all of us think visually, and an important step in making use of the rest of this book is to arrive at some understanding of that process especially as it relates to problem-solving. Visual problem-solving is a method by which we can abstract the elements of a problem so that we can examine it overall and in detail, come to an understanding of its underlying structure and then make judgements required to arrive at a decision. Visual problem-solving is just one form of visual thinking, which is a process of communicating with ourselves that has the advantages of creating a high degree of abstraction, conveying complex relationships, allowing direct comparisons and dealing with the issue of quality. In the following pages we will see how this communication works and how it is used to solve problems.

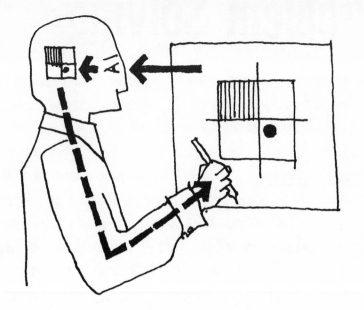

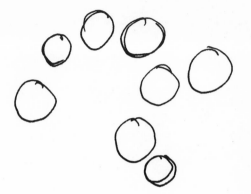

IMAGE ON PAPER

VISUAL THINKING AS A COMMUNICATION PROCESS

Communication with the visual world is a continuing process and so it is with visual thinking. It involves the image, the eye, the brain and the hand. Its potential lies in the exercise of a continuous cycling of information that undergoes transformations at each communication point. We can join the process at any point but will start with the image on the paper (above). Depending upon my background, experience, disposition and my objective in using the diagram I will "see" different things than another person and screen out other things. This is my perceptual image.

18

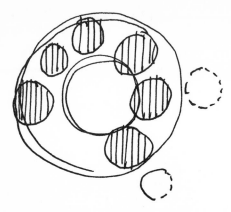 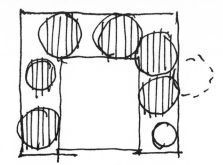 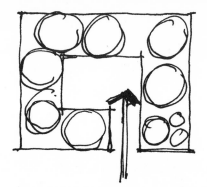

PERCEPTION   MENTAL IMAGE   NEW IMAGE ON PAPER

In this case I am seeking a possible order for the elements of a building so I see a larger circle as the idea of a protected enclave. I have blocked out one of the spaces because it is not important to my perception. Next I form an image in my mind which applies rectilinearity to the perception, probably because it is both a familiar and flexible expression. Finally I set the mental image to paper and the transformation is made as details which were not in my mind are added. This is, of course, a simplification of what is a complex series of many more communications but it should serve as a basic model.

ABSTRACTION

This first step in visual problem-solving — abstraction — cannot be overemphasized, yet it seems to be one of the most difficult things to teach. To understand its function we return to basic communication theory. In any means of communication there is the tendency to communicate things we don't want along with what we do want communicated. Effective communication tries to avoid either giving too much information, thus obscuring our message, or oversimplifying the information so that no message is conveyed. The classic example of this problem is in the design of alphabets. The letters must be sufficiently similar in style but recognizable for their differences. Abstraction is an interaction of the image, the eye _and_ the distance between them. As a rule, images should be small enough to be totally visible without shifting the eyes at the distance which they will be used (for normal desk work smaller than 8½ × 11 sheet), and essential elements must be legible at this same distance. If the image looks too bare, then its size should be reduced to avoid involve-

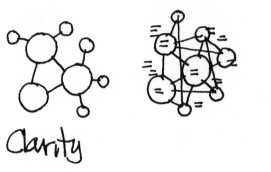

Clarity

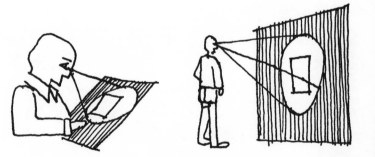

Distance and Size.

20

ment in unessential details. Other concepts which should be considered are hierarchy and conventions. Hierarchy in an image should be familiar to architects. In communication it refers to levels of importance that are assigned to different parts of the message. There are several means for emphasizing those elements which have a higher priority in a visual image. These include centrality, simplicity of shape, intensity or weight, and composition. Size conventions can be very important when communicating with others. The alphabets on the opposite page, for example, retain a measure of legibility because they are familiar symbols arranged in a familiar order. A consistent use of conventional symbols is essential to effective abstraction. Many architects have developed unique styles for their conceptual sketches, their own sort of shorthand. This can sometimes produce certain efficiencies when we are communicating principally with ourselves. However, consistency is still extremely important to the role of abstraction in visual problem-solving.

Hierarchy

Conventions

21

DEVELOPMENT & EXPLORATION

Assuming we now have a fairly clear abstraction of a problem situation, the next step is to manipulate, extend or massage the diagram to 1) discover basic relationships; 2) include supportive information; 3) express alternative configurations of the basic relationships. (This last task will also help clarify the essential nature of relationships.) In the example at right we have a diagram of a typical neighborhood subdivision. Areas such as this threaten to become our next ghettos. The middle class is abandoning such housing for condominiums and P.U.D housing. On the other hand the general condition of the houses makes them ripe for rehabilitation, which may in turn provide an opportunity to solve existing planning problems and create new, more suitable community environments. The problem with the existing set-up is a combination of low-density, wasteful land use and a predominance of automobile rather than pedestrian spaces. The diagrams on the opposite page develop different concepts of available open space. Additional housing units are added in each case and a balance of public and private exterior space is sought.

22

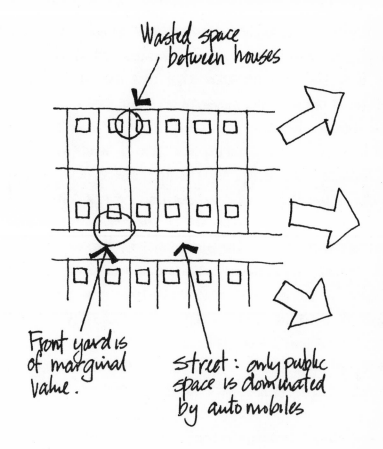

Wasted space between houses

Front yard is of marginal value.

Street: only public space is dominated by automobiles

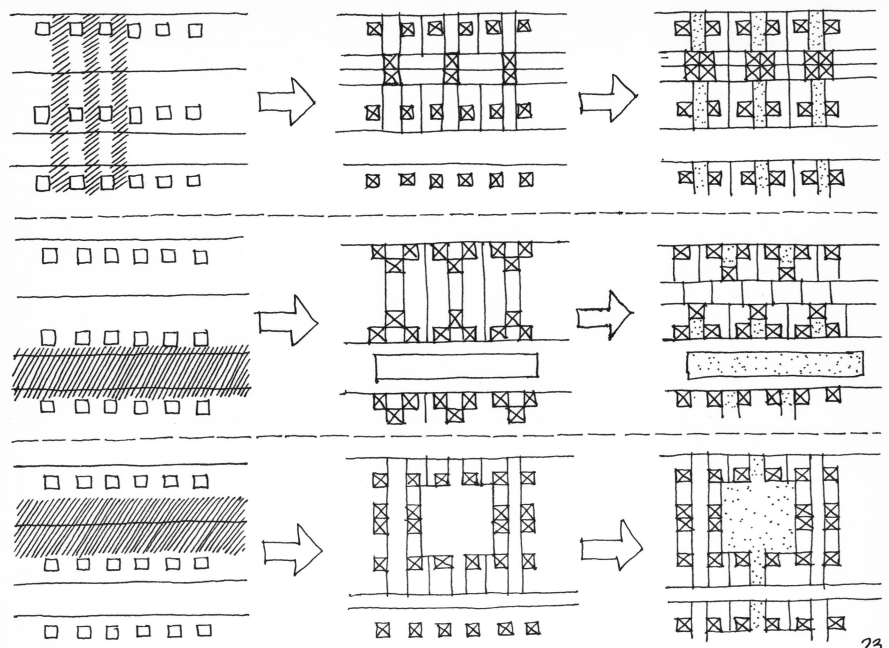

23

MAKING DECISIONS

Problem-solving approaches are not of much value unless they can lead to a point where decisions can be made. The essential ingredients for good decisions are: a set of alternatives from which to choose; a set of evaluation criteria; and a clear way to apply those criteria. In our example the principal evaluation criteria were housing density; community space quality; and effectiveness of the circulation system in supporting pedestrian character. Alternative number ③ was judged to be the best because of the community space that connected twelve houses and because the circulation accommodates the automobile while avoiding conflict between pedestrians and automobiles. An important adjunct to making the choice between alternatives is to adopt the good features from other alternatives in order to help make the decision successful.

The diagrams in the following chapters will deal with abstraction, development, exploration and one or more of the essential ingredients for making decisions. The text accompanying them is intended to relate them to the decision context.

24

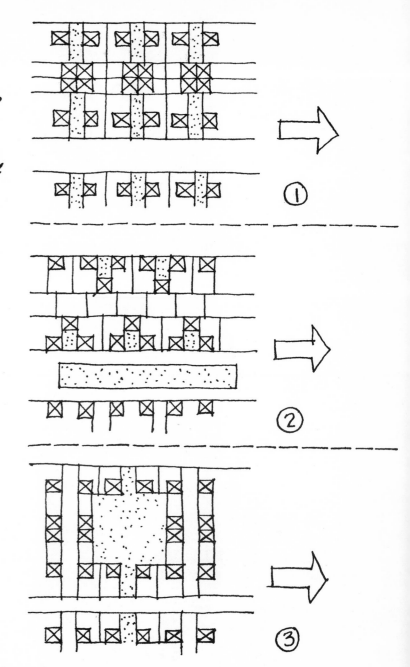

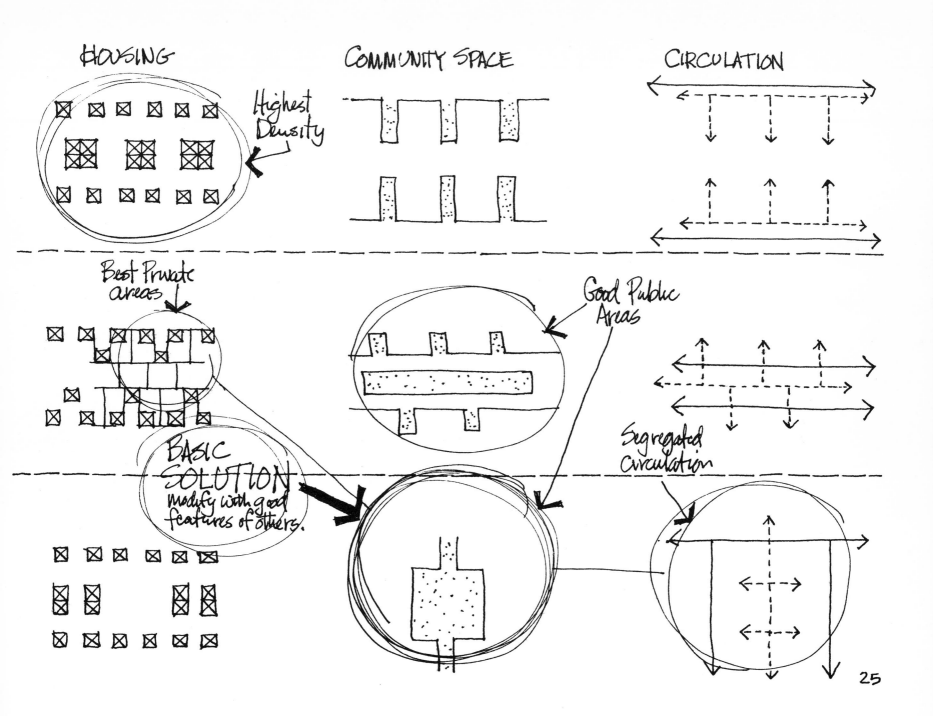

HOUSING

Highest Density

COMMUNITY SPACE

CIRCULATION

Best Private areas

Good Public Areas

BASIC SOLUTION
modify with good features of others.

Segregated Circulation

25

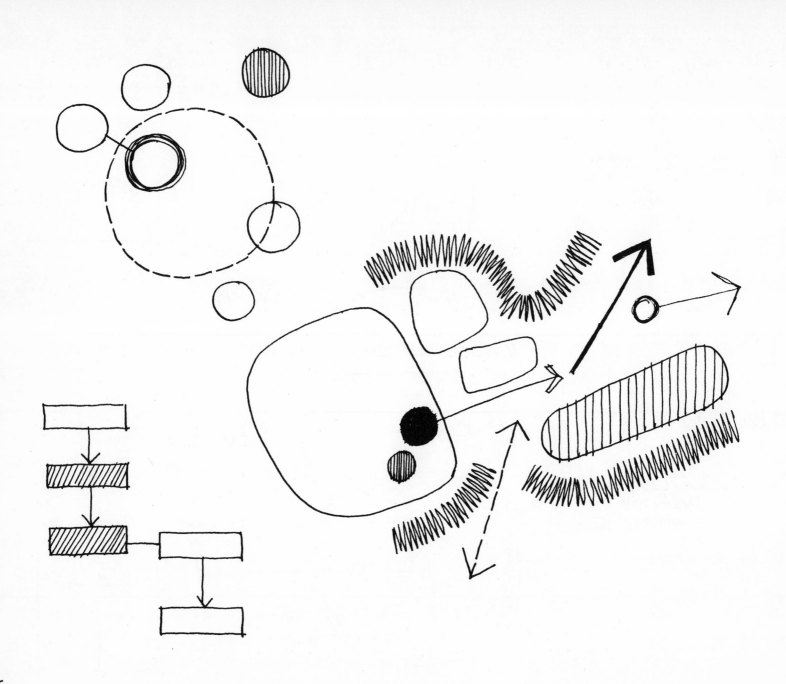

26

Bubble Diagrams

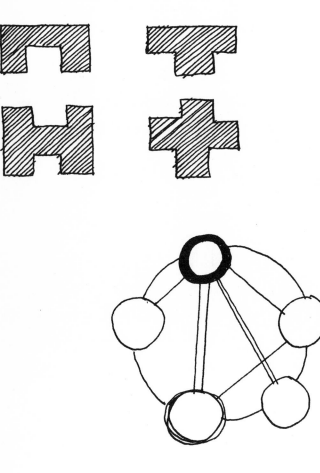

Probably the most versatile and basic device for abstraction, the bubble diagram is useful to the designer or problem solver in a variety of ways. The diagrams may be models of physical space, objects in space, program requirements, activities or existing conditions. The main utility of this form of abstraction is the representation of relationships between a set of elements and the relative ease of manipulating the diagrams to suggest alternative relationships or arrangements. Bubble diagrams are in common use throughout the building industry but with quite a variance in success and often they are not used as effectively as possible. My work with students has indicated some reasons why diagrams are unsuccessful and some directions for improvement. It is important that we see these diagrams as a language which must have the elements of consistency, clarity, and identity required for any successful communication. This chapter will deal first with the issues of language starting with conventional diagrams, then with a wide range of problem types, and finally with problems involving the actual links between bubbles.

27

TRADITIONAL USE : SCHEMATIC DESIGN

Architects have long relied upon visual thinking as a means of arriving at the finished design drawings with which the client or public are familiar. In the example below, the designer of a house seeks the best over all position of the rooms with respect to the site. Before he begins to make decisions, the designer diagrams all the required

spaces, showing the necessary relationships. He draws an abstract diagram which purposely does **not** establish the size and shape of the rooms or their position on the site (A). In the second diagram (B), the position of all the rooms is established. The orientation of the site, direction of views and site or building access will have been considered.

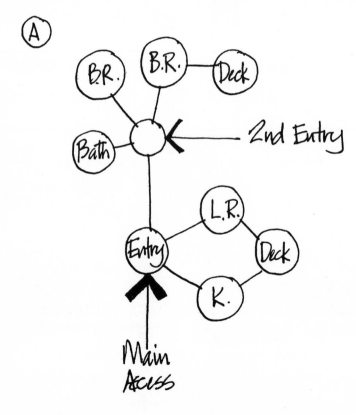

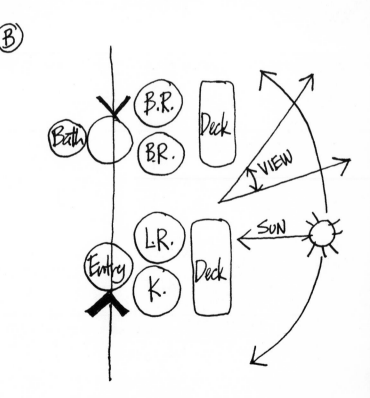

28

The next step, Ⓒ, focuses on the size and shape of the rooms. The furniture and activities within each room and the continuity of the spatial experience become important at this point. Note: The scale of the diagrams should probably increase in size and should be a specific, standard scale. The final diagram, Ⓓ, tightens the plan by making assumptions about construction, structure and enclosure. We are now close to the point where hardline drawings can be started. This last drawing can also be used as a means to select and identify required details and sections. The bubble diagrams have forced the designer to proceed from the general to the particular.

Ⓒ

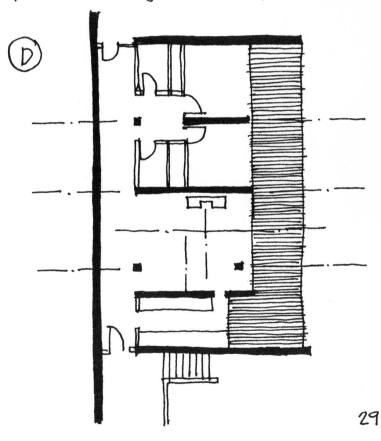

Ⓓ

29

Ⓐ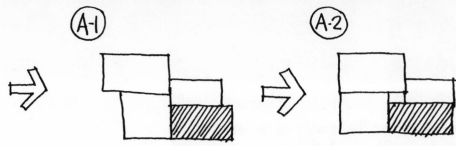

Ⓐ-1 Ⓐ-2

Ⓑ

Ⓒ

Ⓓ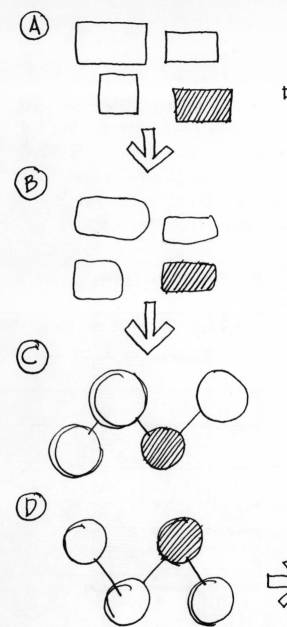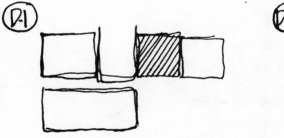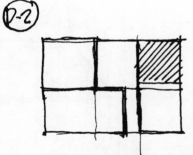

Ⓓ-1 Ⓓ-2

30

DIAGRAM LANGUAGE : ABSTRACTION

The most fundamental control that must be exercised in using diagrams is the level of abstraction. In the sketches (A), (A-1), (A-2), we can see the difficulties that arise when shapes become too specific and are taken literally. By contrast sketches (B) and (C) increase the degree of abstraction, allowing for a looser interpretation of the diagram and thus a freer exploration of solutions in sketches (D), (D-1), (D-2).

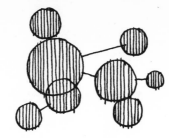

DIAGRAM LANGUAGE : GRAMMAR

On the right are shown three ways to organize the graphic elements of zone, point and line. They are: free form, where the positions of the zones are arbitrary; structured, where an implied or explicit grid lends significance to the position of zones; profile, where the shape or size of zones seen comparatively convey another level of information.

Another type of organization of bubble diagrams is the three-dimensional, such as in the example (E). Although organization alternatives are similar to those for two-dimensional diagrams, the third dimension allows more flexibility in showing relationships of zones or points.

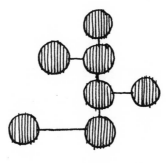

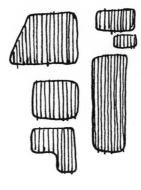

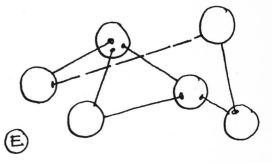

(E)

31

DIAGRAM LANGUAGE: THE ELEMENTS

ZONE

POINT

LINE

Basic Diagrams

Contrary to what many people think, bubble diagrams are not limited in their use to very complex problems. These diagrams can be useful on simple, everyday problems. It is perhaps here that the contribution of visual thinking can be most clearly seen. For example: If we break down the family budget into bubbles sized to match percentages of the budget, we can already see the relative importance of budget items. By distinguishing controllable from uncontrollable budget items, the bubbles reveal the places where expenditure adjustments might be considered.

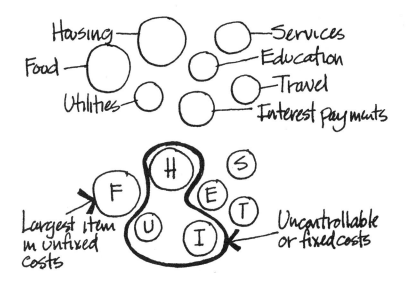

Housing — Services
Food — Education
Utilities — Travel — Interest Payments

Largest item in unfixed costs

Uncontrollable or fixed costs

The fancy name for this approach to problems is establishment of problem boundaries. Visually segregating those things which we can't change helps us to focus on what we can change while keeping us aware of the relationships between both.

Below, this approach is applied to a typical project situation in the initial stages of conversation between client and architect. Problem boundaries have been drawn in different ways to illustrate the range of services an architect might render.

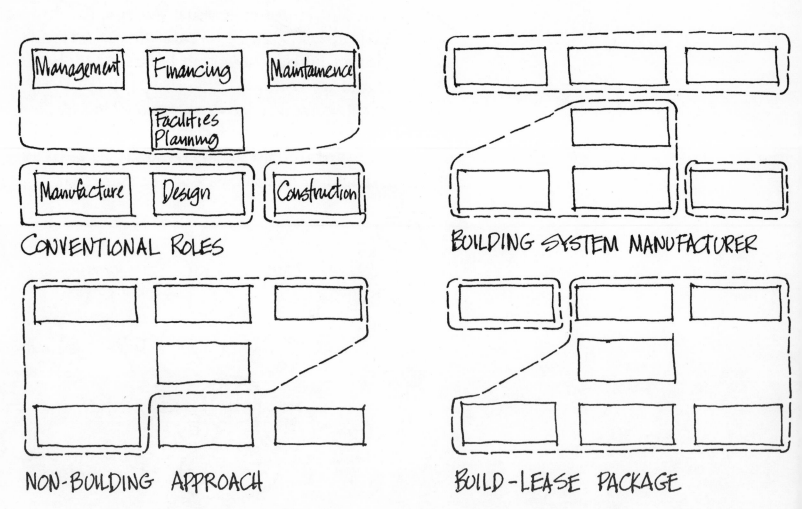

CONVENTIONAL ROLES

BUILDING SYSTEM MANUFACTURER

NON-BUILDING APPROACH

BUILD-LEASE PACKAGE

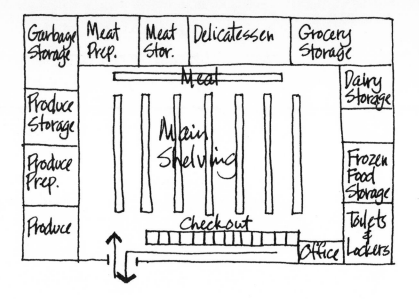

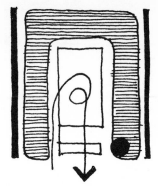

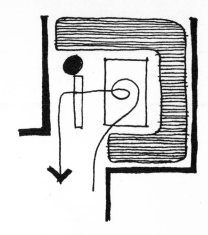

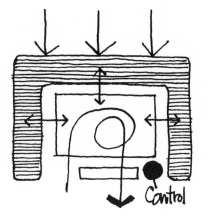

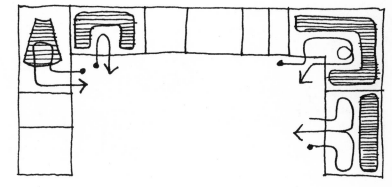

CONDENSATION OF PROGRAM INFORMATION

Often the client's program is given to us in the form of standard layouts that meet the program requirements, as in the case of a supermarket above. I sometimes find it helpful to make an abstract diagram that reduces the information to a visual level, which is easier to manipulate and to remember. This diagram can then be adjusted to check the implications of different site constraints. When several stores, restaurants or theatres must be planned jointly, these simplified diagrams may help.

35

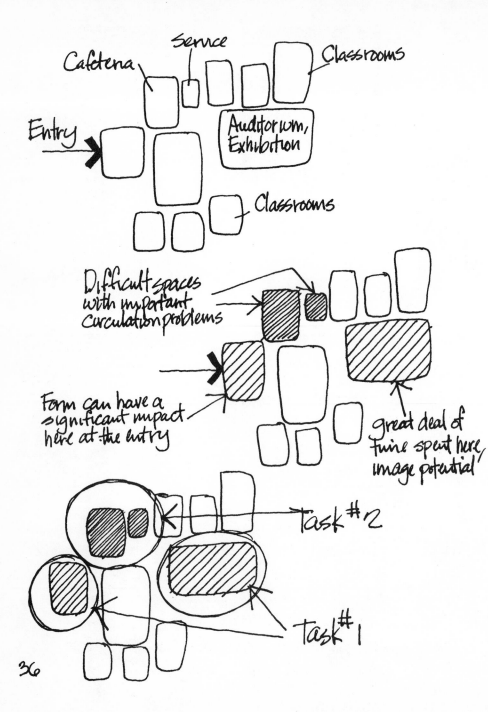

Cafeteria

Service

Classrooms

Entry

Auditorium, Exhibition

Classrooms

Difficult spaces with important circulation problems

Form can have a significant impact here at the entry

great deal of time spent here, image potential

Task #2

Task #1

36

SETTING DESIGNING PRIORITIES

With the tightening of building design budgets we are becoming aware of the need to concentrate efforts where they will have the greatest impact on the quality of design. Diagrams made at the conceptual stage may serve to clarify critical design problems and to act as a visual reminder of design objectives throughout the project.

HIERARCHY OF SPACES

The diagrams at right show the underlying rationale for the relationships of spaces in a college dormitory building. Most important, they show the hierarchical organization of spaces without restraining the alternatives for physical design solutions. This way we can get a look at the program as a determinant without having it override the design.

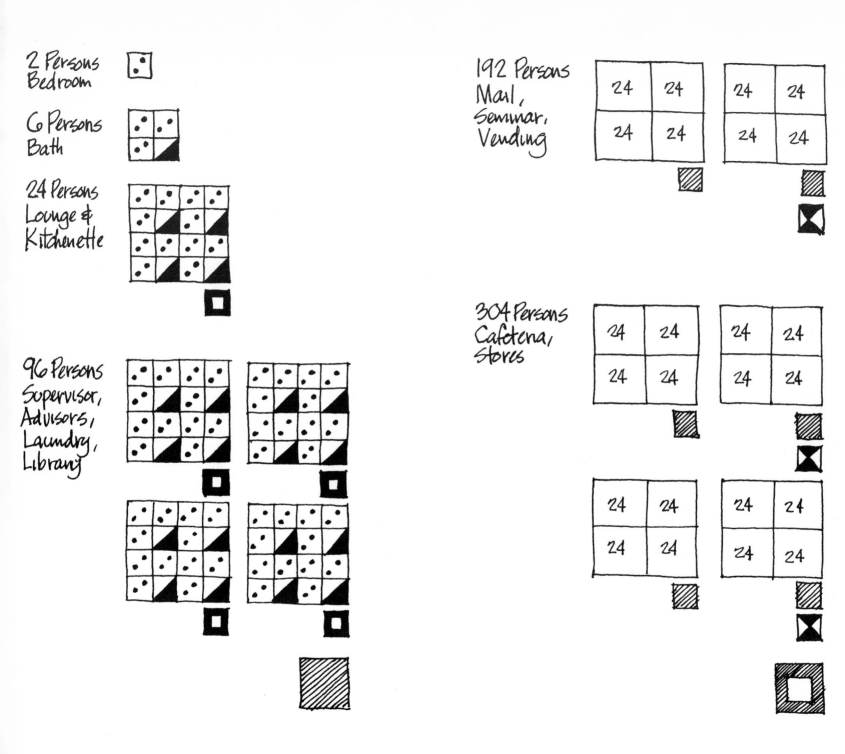

2 Persons
Bedroom

6 Persons
Bath

24 Persons
Lounge &
Kitchenette

96 Persons
Supervisor,
Advisors,
Laundry,
Library

192 Persons
Mail,
Seminar,
Vending

| 24 | 24 | 24 | 24 |
| 24 | 24 | 24 | 24 |

304 Persons
Cafeteria,
Stores

| 24 | 24 | 24 | 24 |
| 24 | 24 | 24 | 24 |

| 24 | 24 | 24 | 24 |
| 24 | 24 | 24 | 24 |

37

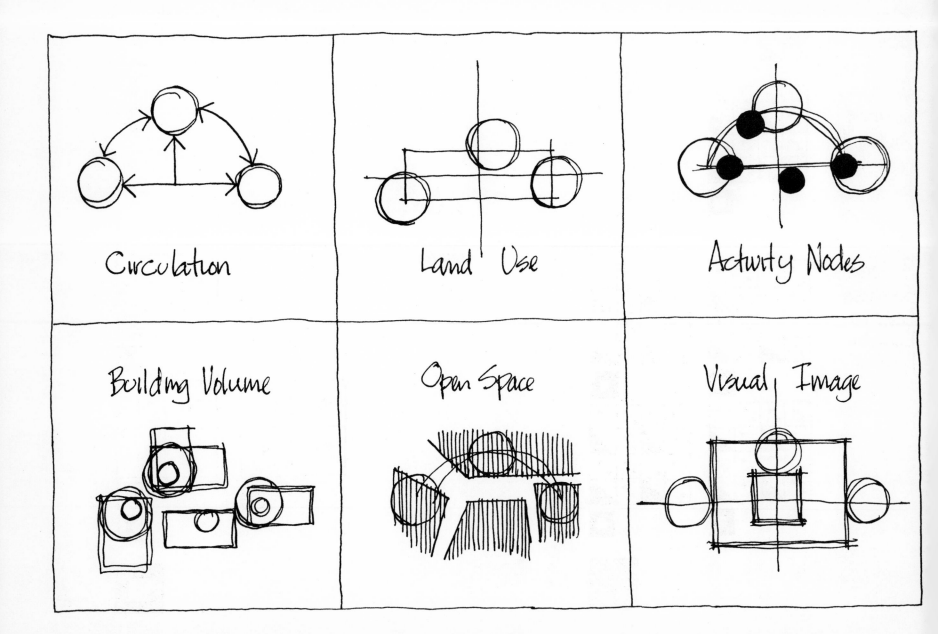

Circulation

Land Use

Activity Nodes

Building Volume

Open Space

Visual Image

DESIGN DETERMINANTS

Some architects find that visualizing design determinants helps in the generation of design solutions. The diagrams at the left were generated for a shopping mall. Their value lies in the variety of perceptions which can be formed by further review of the images.

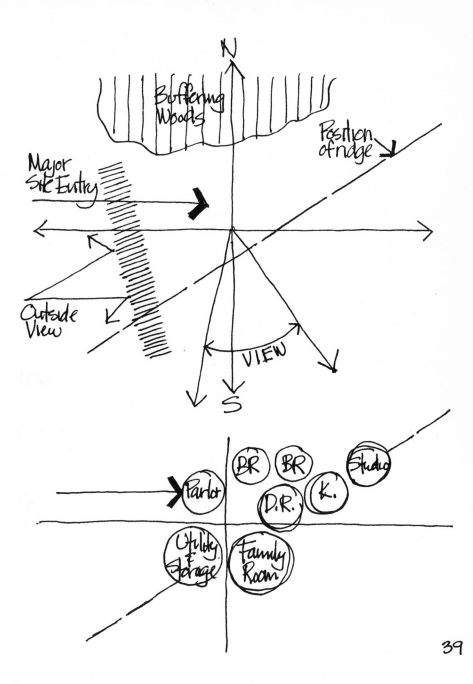

SITE CONSIDERATIONS

In developing a preliminary model for the design of a house on the right, a diagram is made to show several site conditions at the same time. Orientations which should be screened out are shown as well as preferred orientations such as sun and view. In the second diagram, ideal locations for rooms are shown. The solution alternatives are left open. Adjustments might be made to accommodate all rooms on one floor, or the rooms might be vertically shifted or stacked depending upon the terrain and approach to construction.

Airport Layout
Basic Types

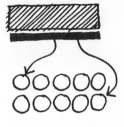

Combinations

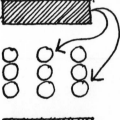

Variations

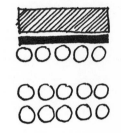

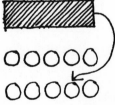

LAYOUT ALTERNATIVES
By abstracting alternatives to the level of their basic differences quick reference can be obtained. This is a big help when struggling with initial concepts. On the left, basic choices are summarized for airport layouts.

SITE ORGANIZATION
Below alternative concepts are shown in simple diagram form. This type of quick sketch can be of help in thinking about the way in which the contractor might set up storage, locate trade shops and handle access and traffic.

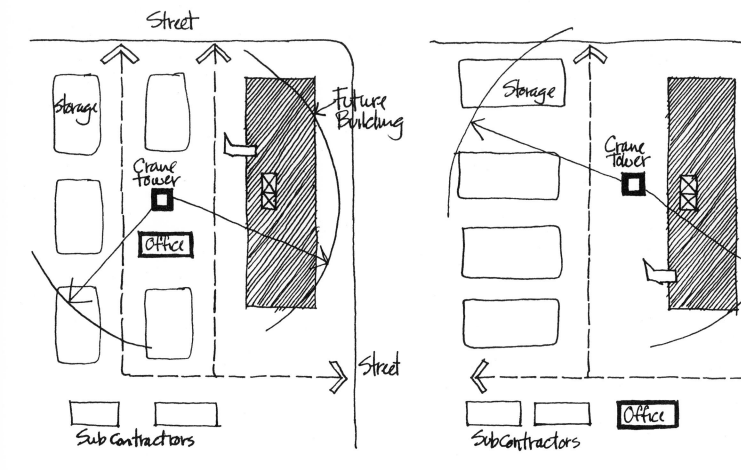

Street

Storage

Crane Tower

Office

Future Building

Street

Sub Contractors

Storage

Crane Tower

Office

Sub Contractors

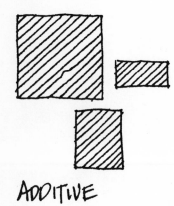

ADDITIVE

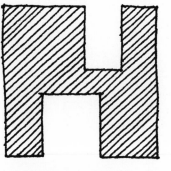

SUBTRACTIVE

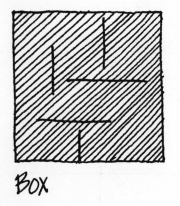

BOX

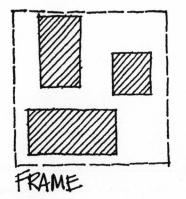

FRAME

BUILDING TYPES

The sketches on the left should be familiar. They represent four different ways of considering the form of a building. These abstractions can be used to generate several plans for a specific building. They are economical images and can be easily held in our memory.

STREET PATTERNS

In a similar way it is possible to sum up primary options for planning a housing development as on the opposite page.

42

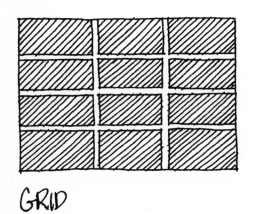

GRID

LOOP

LOOP & TEE

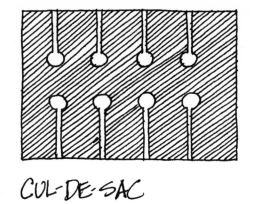

CUL-DE-SAC

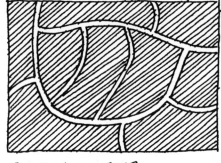

CORVILINEAR

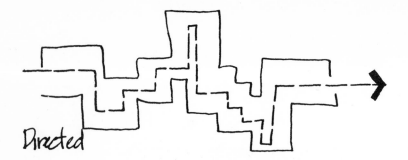

Directed

CIRCULATION TYPES
One way of stating design objectives is the visual description of the desired behavior to be facilitated by the building layout. Here we have four possible objectives with regard to circulation of the public within an exhibition.

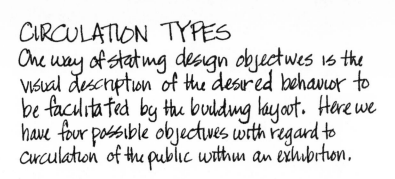

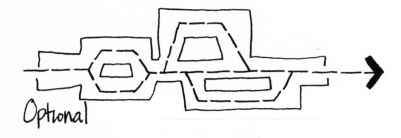

Optional

Non Directed

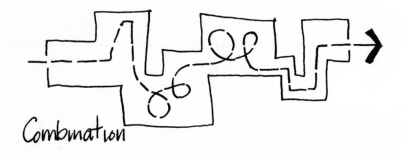

Combination

Links

As was pointed out at the beginning of this book, one important advantage of graphic problem-solving is the capability of focusing upon the relationships within an organization of parts, a system or a problem. It is possible in many instances to understand the basic characteristics of a situation or an operation by focusing on the distinguishing features of its links or connections. Once we shift our attention from the bubbles in a diagram to the lines or links between bubbles, another set of uses for these diagrams, as shown in this section, becomes apparent. It is important to us here to use graphic means to reinforce the emphasis upon links. One means is the use of a variety of expressions for lines such as on page 32. A second subtle but equally effective means is the neutralizing of the bubbles in their shape or position.

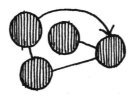

Equal size and shape

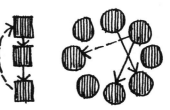

Equal position

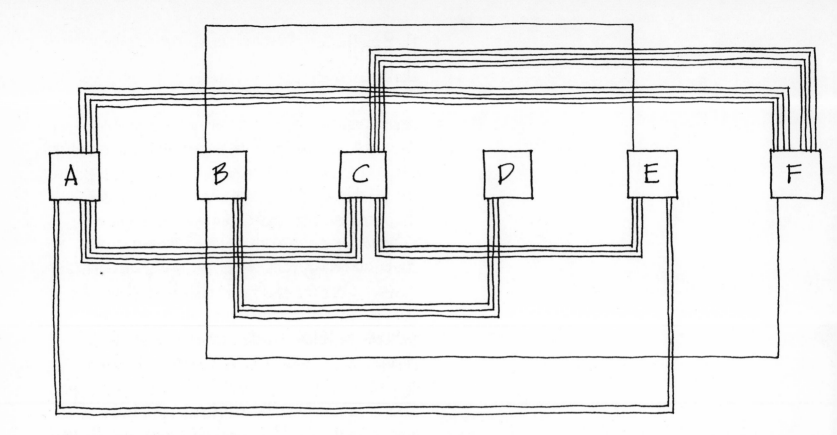

INTENSITY OF RELATIONSHIPS

A common form of link articulation is that which helps to explain the intensity of interaction desired between programmed activities. Here only one of several approaches is shown. Initially the activities are represented by boxes arbitrarily arranged in a line. Intensity of relationships is indicated by the number of lines linking the boxes. The next step tries to locate intense relationships closer to each other. The third step tries to organize the diagram for clarity. The fourth step explores the possibility of indicating intensity by proximity of the bubbles. Finally both proximity and the thickness of the links express intensity.

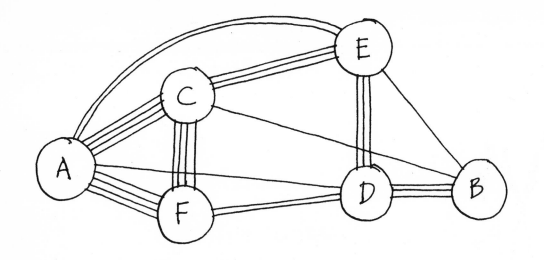

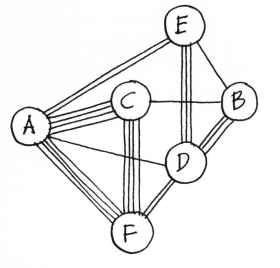

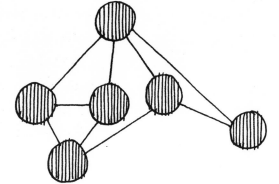

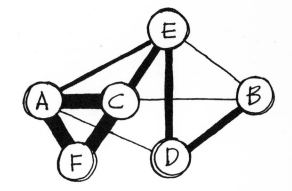

47

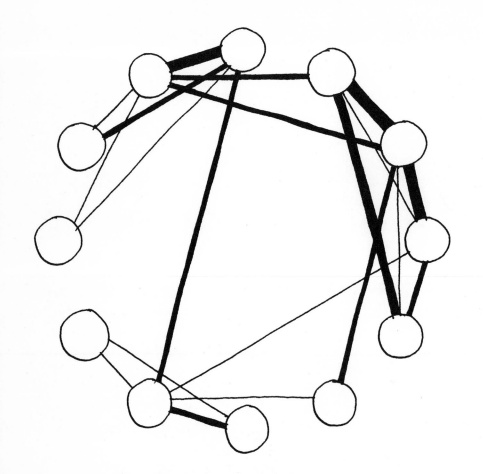

The diagram at left shows another way to indicate intensity of relationships. It has the advantage also of being able to compare the relationships as a group with the spaces as a group.

PROBLEM PRIORITIES
In designing certain spaces we usually must address several related problems. The solution to certain problems will often affect the solution of other problems. To the designer who is interested in finding those key problems the dependency diagram at right should help. The position of the dots, which represent the problems, is arbitrary; the arrows indicate the direction of dependence.

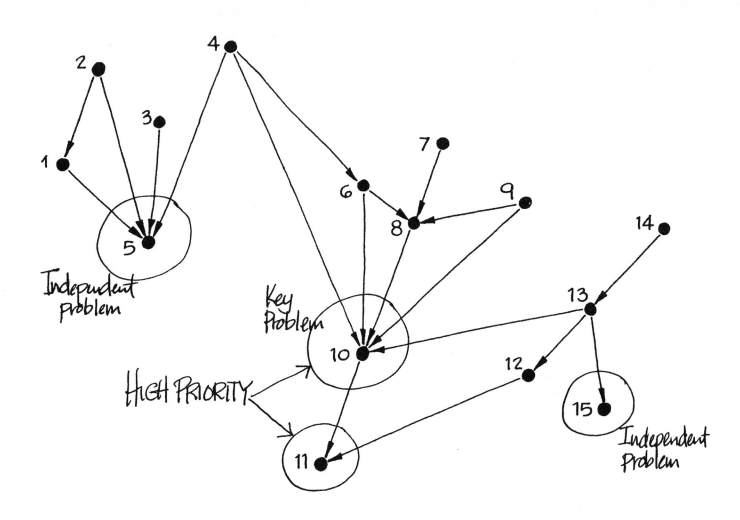

Independent
Problem

Key
Problem

HIGH PRIORITY

Independent
Problem

49

CONCEPT GENERATION

The contrast between ideal circulation and realistic site-related circulation can be the source of initial building concepts. By reducing the visual information and overlapping the circulation types in the same diagram and then considering vertical circulation, preferred positions for stores or exterior spaces as in a shopping center can be identified. In the final diagram the previous strong expression of ideal circulation suggests a formal expression.

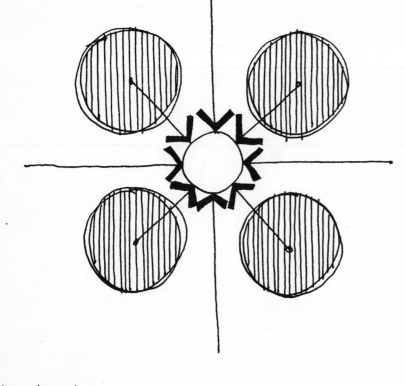

Abstracted ideal access

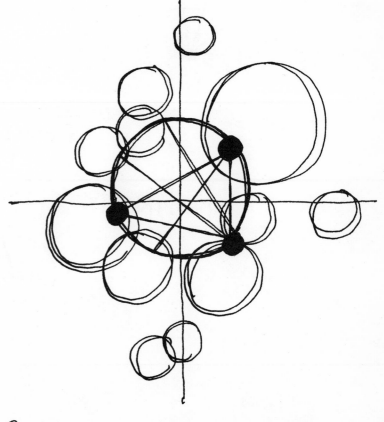

Relative sizes & Major Circulation

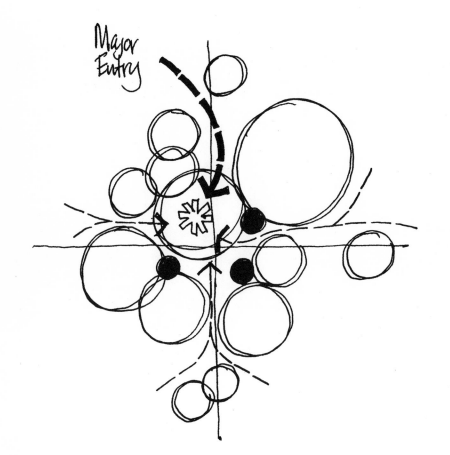

Major
Entry

Resolution & Basic Concept

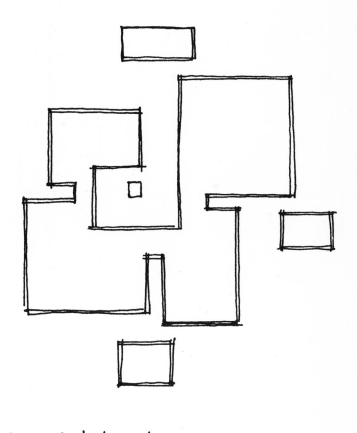

Formal Alternative

51

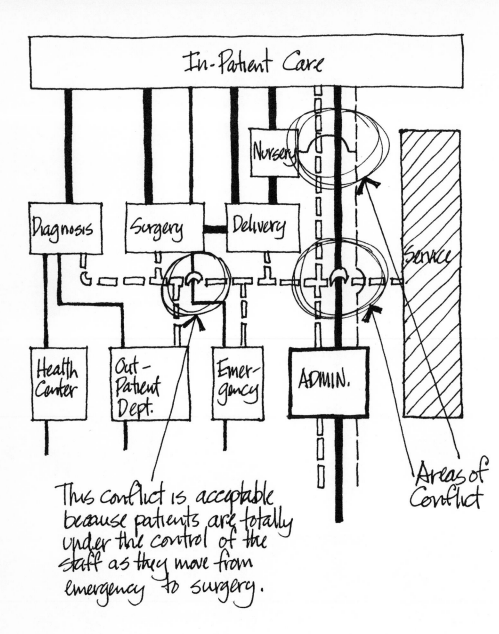

In-Patient Care

Diagnosis

Surgery

Delivery

Nursery

Service

Health Center

Out-Patient Dept.

Emer-gency

ADMIN.

Areas of Conflict

This conflict is acceptable because patients are totally under the control of the staff as they move from emergency to surgery.

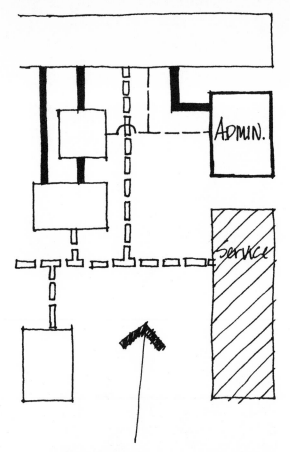

ADMIN.

Service

REDUCTION OF CONFLICT by switching administration and service.

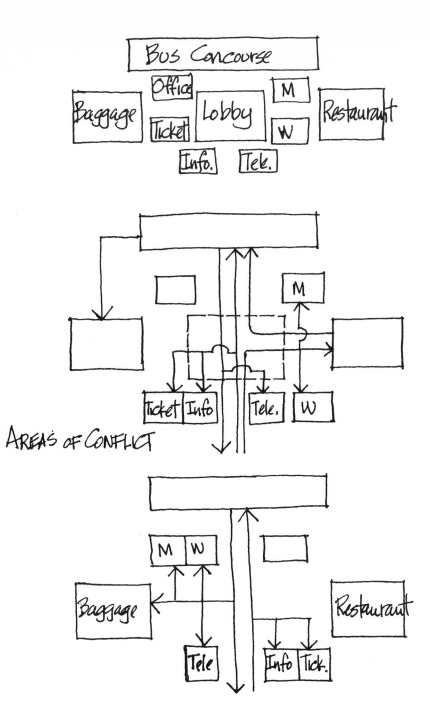

AREAS OF CONFLICT

RESOLVING CIRCULATION CONFLICTS

It is common practice in projects with complex circulation requirements, such as a hospital, to illustrate the program of spaces as an organization diagram prior to conceptual design work. There are two important graphic considerations in this type of application: 1) clear distinction of traffic types. Maximum contrast between line expression should be maintained and there should be a generous amount of space between any parallel lines; 2) points of circulation conflict or crossover should be highlighted. I prefer the standard electrical symbol for this because it makes obvious the conflicts that must be resolved.

The first example on the opposite page deals with patient, staff and visitor circulation in a typical hospital. The second example is a study of a standard layout for bus terminals. Many of the approaches to this diagram type will also be useful when we deal with activity networks in the last section of this book.

ENVIRONMENTAL CONTROL

With the emergence of energy Conservation as a Key issue in building, it is necessary for the designer to bring mechanical systems to the same level of conceptualization as traditional "architectural" design issues. We need to grasp the basics of environmental control and the emerg-ing engineering solutions. The visual tools presently in use are bubble diagrams such as are shown on these pages. Below a system for maximum use of energy is summarized. On the opposite page are two examples where alterna-tive approaches to distribution are compared. Note the importance of symbolic shapes.

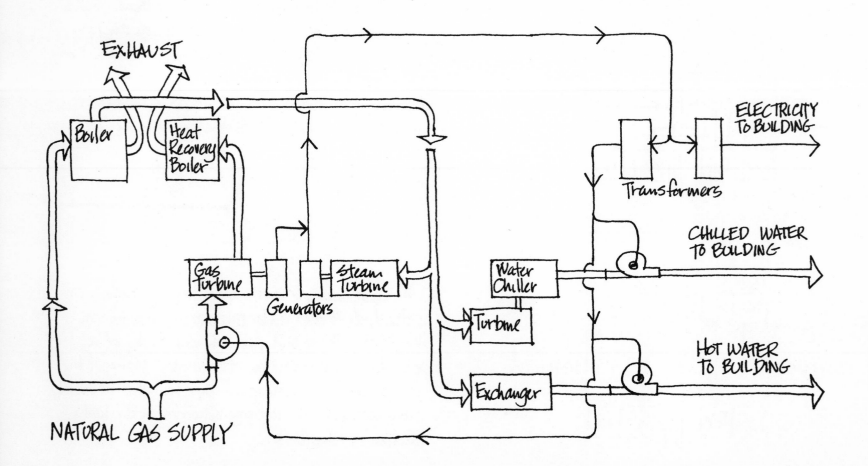

EXHAUST

Boiler

Heat Recovery Boiler

Gas Turbine

Generators

Steam Turbine

Water Chiller

Turbine

Exchanger

Transformers

ELECTRICITY TO BUILDING

CHILLED WATER TO BUILDING

HOT WATER TO BUILDING

NATURAL GAS SUPPLY

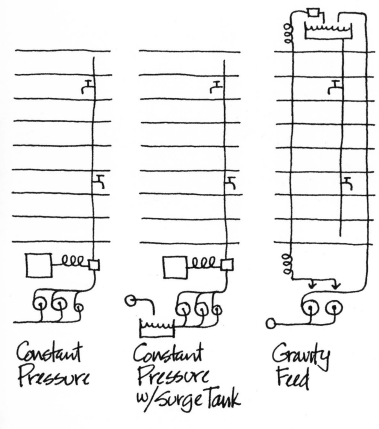

Constant
Pressure

Constant
Pressure
w/Surge Tank

Gravity
Feed

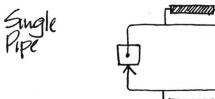

Single
Pipe

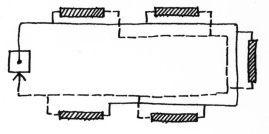

Two Pipe
Series

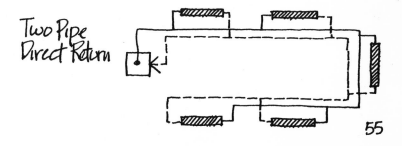

Two Pipe
Direct Return

55

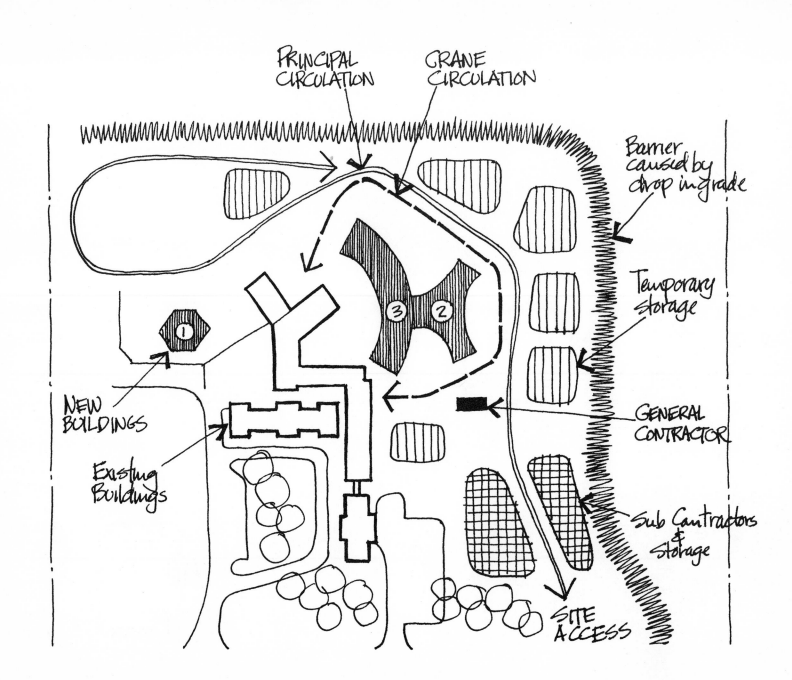

PRINCIPAL
CIRCULATION

CRANE
CIRCULATION

Barrier
caused by
drop in grade

Temporary
storage

GENERAL
CONTRACTOR

NEW
BUILDINGS

Existing
Buildings

Sub Contractors
& Storage

SITE
ACCESS

56

CONSTRUCTION SITE CIRCULATION

Often in projects such as a hospital expansion phasing of work and site organization are critical to minimum interference with ongoing use of the existing buildings. Because of the large spread of the new buildings a moving crane is necessary. Its circulation path as well as the general circulation form the basis for site organization. Buildings are generally to one side of the circulation with temporary storage to the other side. The general contractor's office is placed in a sort of traffic control position.

PERSONNEL ORGANIZATION

Effective use of people and their skills is essential to constructing good buildings. The organization chart at right shows one way of coordinating engineering with the management functions; it helps to clarify roles and responsibilities of individuals.

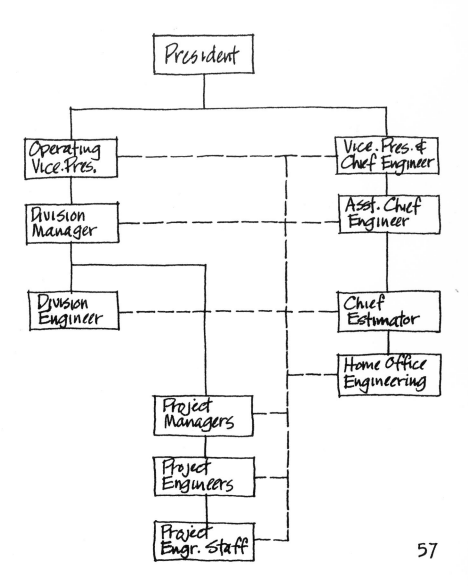

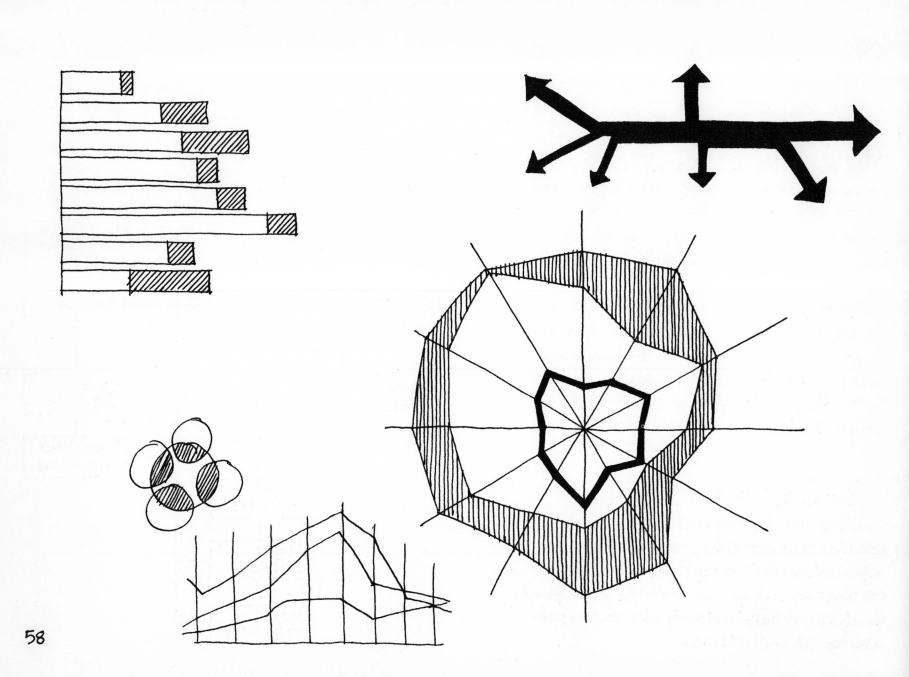

58

Area Diagrams

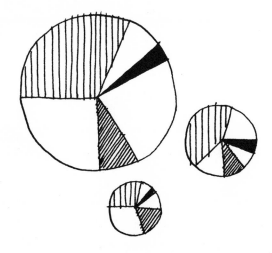

The diagrams in this section differ from simple bubble diagrams in one fundamental aspect. Each diagram has an assumed scale that allows us to read relative sizes, be they areas, lengths or widths. The diagram's value lies in its capability to summarize in a way that allows quick comparison or reference. After reviewing a variety of applications of simple area diagrams, examples that deal with issues of intensity are shown. Finally the element of time is applied to quantitative measures exploring trends and forecasting.

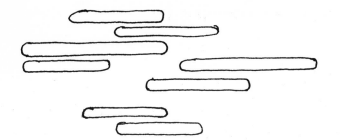

One special note: Scale is subject to distortion in diagrams in the same way as statistics, particularly when two scales are used in the same diagram. The communication of quantities is often conditioned by general conceptions of scale. This should not be a problem as long as we always check that the graphics and the content match.

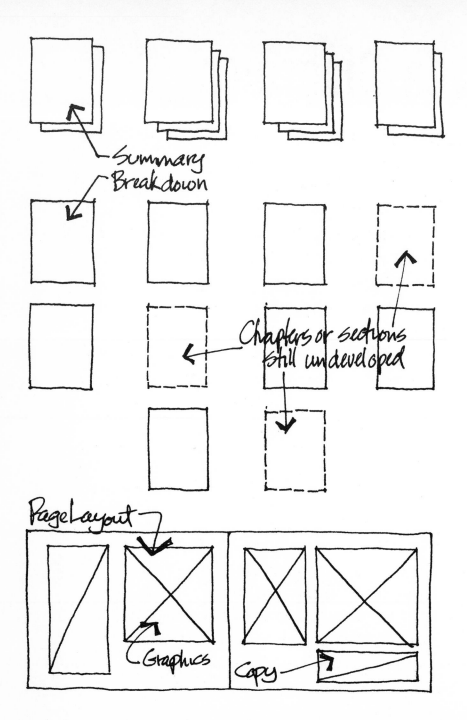

Summary
Breakdown

Chapters or sections
Still undeveloped

Page Layout

Graphics

Copy

ORGANIZING REPORTS

With the increasing complexity of building pro-grams and the need for research, written reports for client or internal use have become more common. Graphic communication is an accepted part of building design presentations, but curi-ously when we have to produce a report, many of us leave our graphic sense behind. A report is also a visual tool, even if a good deal of its language is written. The sequential nature of the written statement makes the organization and the accessibility of a report critical objec-tives. Graphics can be an important tool in this regard. Two useful devices are the visual summary of the elements of the report and the mock-up of specific pages, both of which were used in preparing this book.

PROGRAM ANALYSIS

On the right is an elementary but often ignored device for getting a visual sense of the design implications of space requirements. I prefer the use of squares because they help in reading the relative size of spaces without implying a parti-cular configuration of the individual spaces and because the square is easy to calculate.

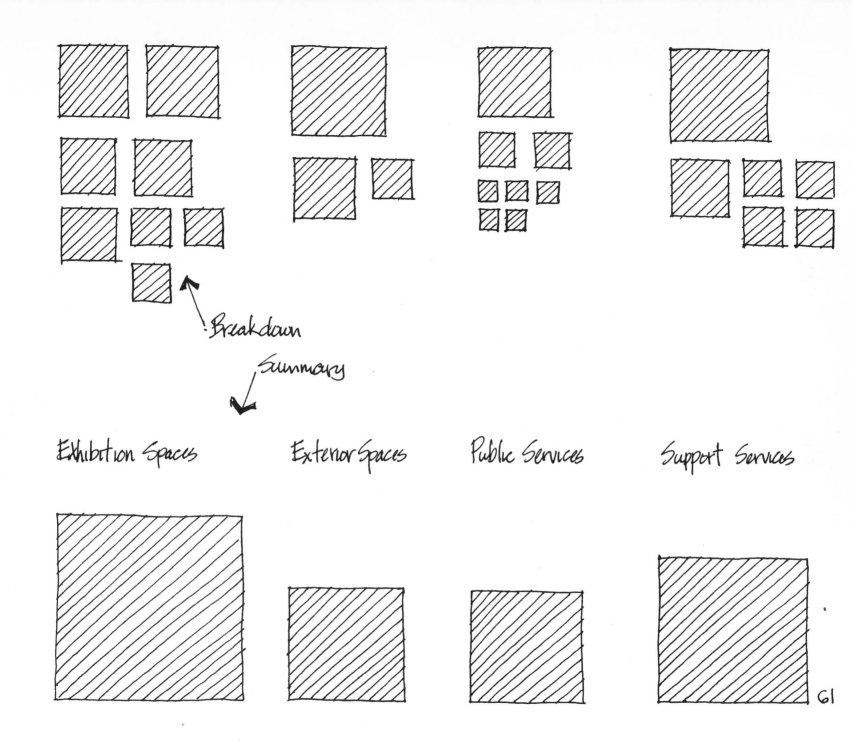

Breakdown

Summary

Exhibition Spaces   Exterior Spaces   Public Services   Support Services

61

Available Land

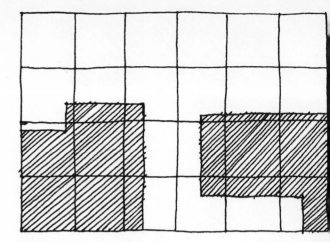

Zoning

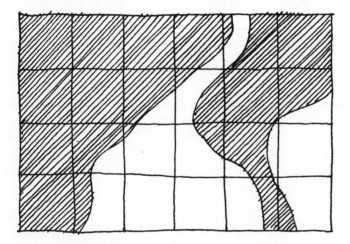

Geological Criteria

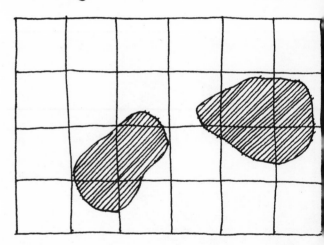

Prime Building Conditions

62

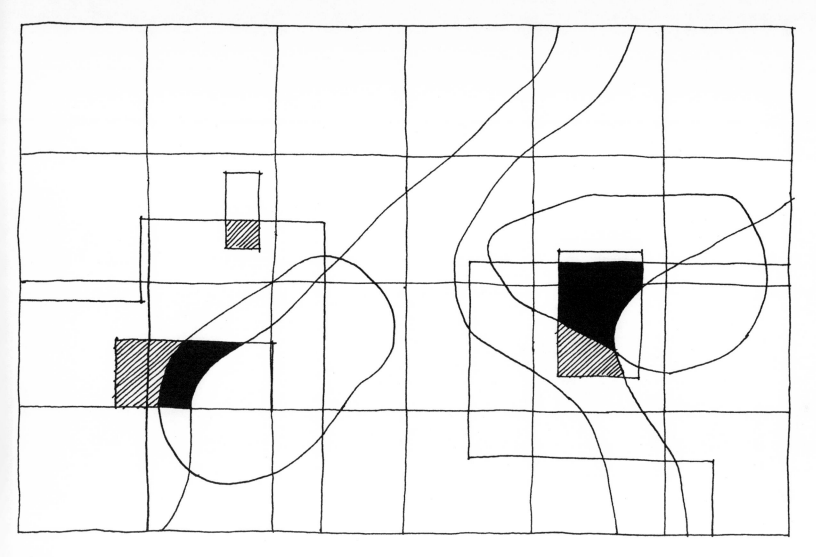

SITE SELECTION

This is a graphic approach to help simplify the analysis required for site selection; at the same time simplify the explanation for the client. A large number of considerations ranging from land costs to subsoil conditions are arranged within a few categories. Favorable sites are illustrated for each category. By laying the diagrams over each other the ideal sites can be located. Possible second choices appear where three categories overlap.

63

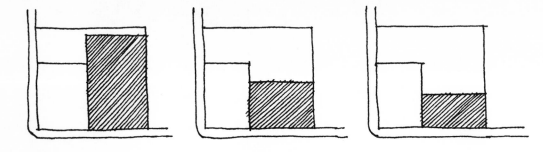

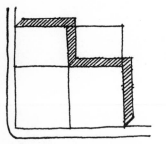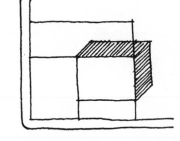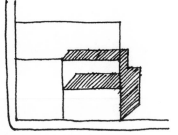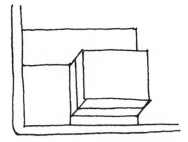

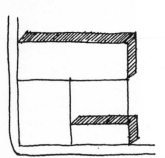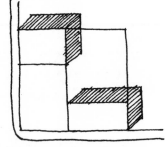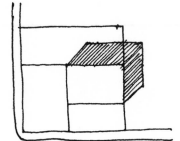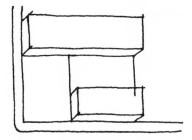

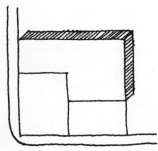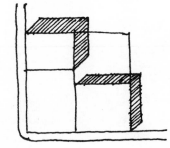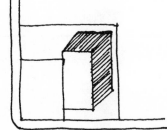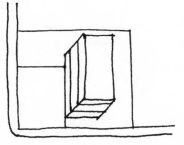

BUILDING MASSING

The thumbnail sketch is often overlooked as a means to generate a range of options for the relationship of building mass and the site. Of course these sketches are only symbols for several factors which would affect a final decision.

MASSING AND IMAGE

Area diagrams can also be helpful in showing abstract attitudes toward the basic posture of a building. These in turn can quickly generate a variety of building shapes. The diagrams on the right show just a few of the possible inter-pretations. As the site becomes more complex or program requirements take priority, the number of alternative shapes increases. Meanwhile the original diagrams are clear reminders of the basic postures.

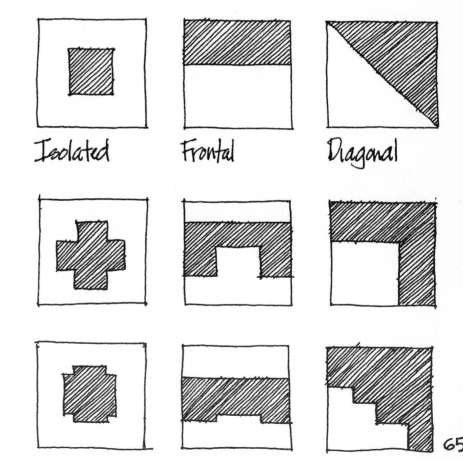

Isolated   Frontal   Diagonal

65

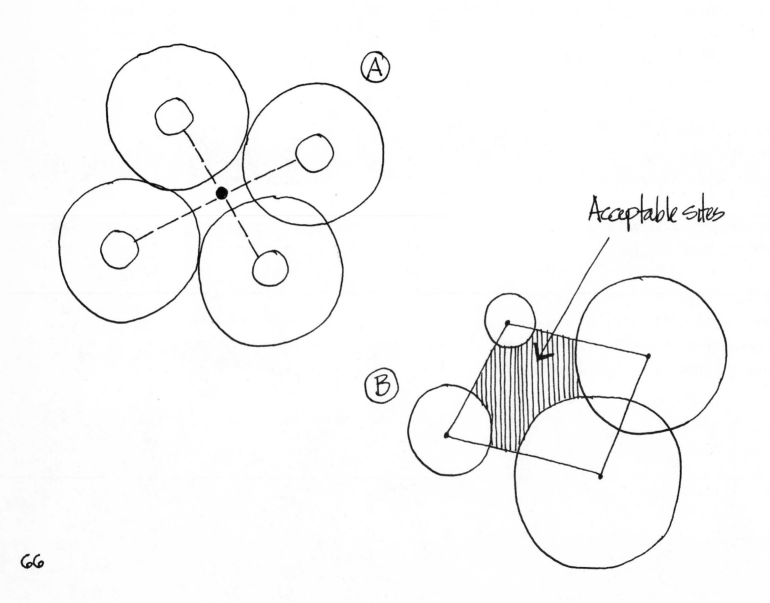

A

B

Acceptable sites

FACILITY LOCATION

Assume that we are considering the location of a new commercial facility or fabricating plant that will serve four market areas. The facility could be placed in the geographical center (A). However, if markets are considered to be of unequal importance, the most important market can be shown as the smallest circle with lesser markets of larger size circles (B). In this way the diagram can be used to indicate a general zone of acceptable sites for a facility. This does not take into account several other factors such as available transportation and condition of routes, but it does provide a starting point for further decisions.

Another way to deal with spheres of influence is in terms of actual travel time (C). In this case we can establish different zones of service for an emergency community facility.

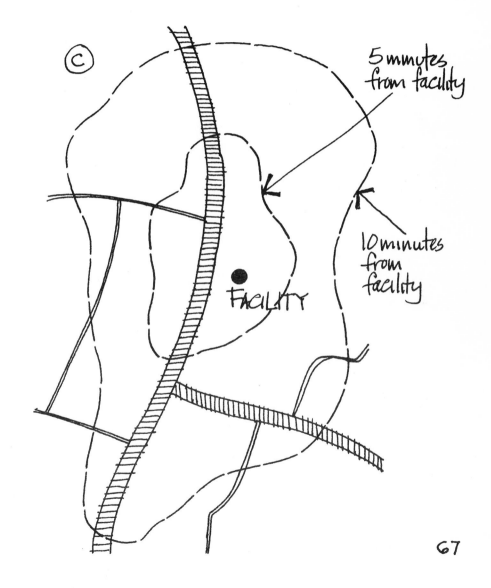

(C)

5 minutes from facility

10 minutes from facility

FACILITY

67

SITE, CONTEXT ANALYSIS

The reduction of existing elements of a site
and its context to simple graphic abstraction
can help in grasping basic design opportunities.
The identity of different parts is extremely
important in this type of diagram. I try to
choose symbols for zones, lines and points that
have simplicity and sufficient contrast to be
easily read.

Intensity Mapping

A very important part of making design decisions is the setting of priorities. The needs of people cannot be plugged into some formula, but we can at least identify which needs we are considering and make a value judgement about which are the most important.

    The mapping of intensities can be helpful in setting priorities and keeping them in view. A simple illustration of this is a diagram of kitchen usage in preparing a meal.

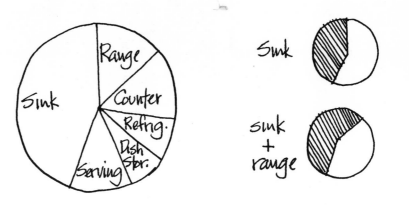

The percentage distribution of trips when shown as segments of a circle can be easily read and recalled.

    The diagrams on the following pages show a wide range of alternatives for this form of sketch.

FACILITY USE

The bar chart immediately below shows the use of several different facilities by different age groups. It is possible to see the aggregate use of facilities as well. This gives a quick comparative look at the volume and characteristics of facility demand. The other chart shows another way to diagram the simultaneous use of three facilities over time. The bar charts on the opposite page show the degree of use for four different facilities during a typical day. Each chart is condensed at the far right to make the patterns of use easier to compare or contrast. These smaller diagrams also dramatize the differences in the total aggregate use.

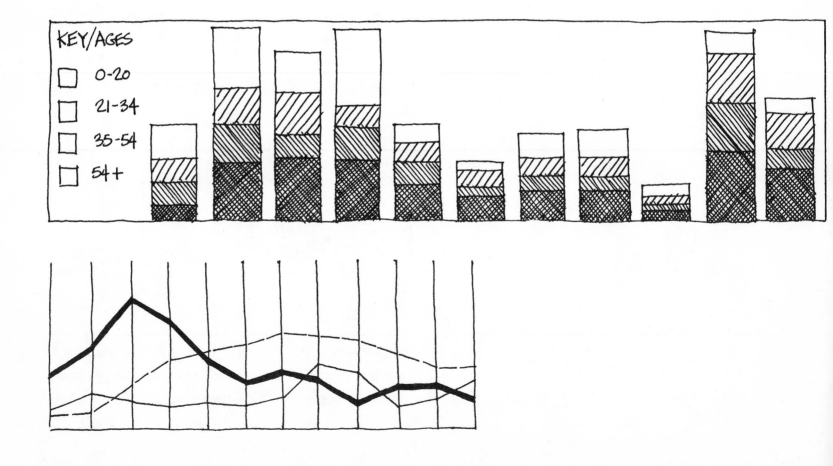

KEY/AGES

☐ 0-20

☐ 21-34

☐ 35-54

☐ 54+

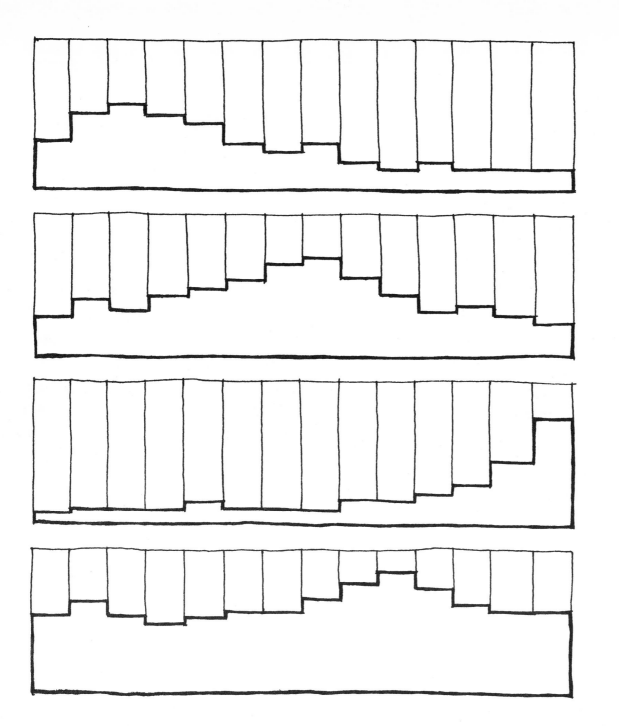

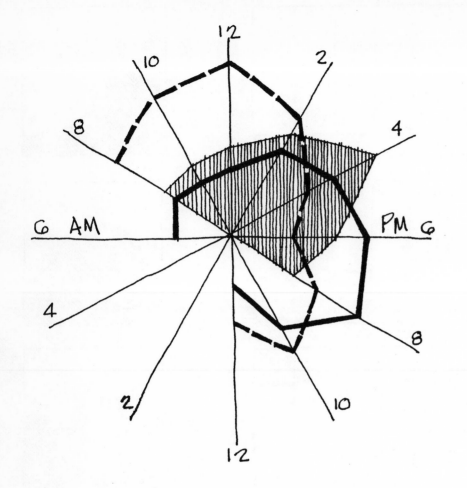

RADIAL CHARTS

These are alternate means of diagramming linear or bar chart information. Their effectiveness lies in certain conventions with which most of us are familiar. On this page a 24-hour clock similar to a sun dial is used to show intensity of use for three different facilities.

On the opposite page four different evaluation criteria are assigned to four scaled vectors. The ratings are marked for each criterion (1-5) and connected. The area of the resulting quadrangle represents the total score. Balanced scores among criteria create larger areas Ⓐ. Different numbers of criteria can be used Ⓑ, but I prefer the quadrangle when possible because of the ease of calculations.

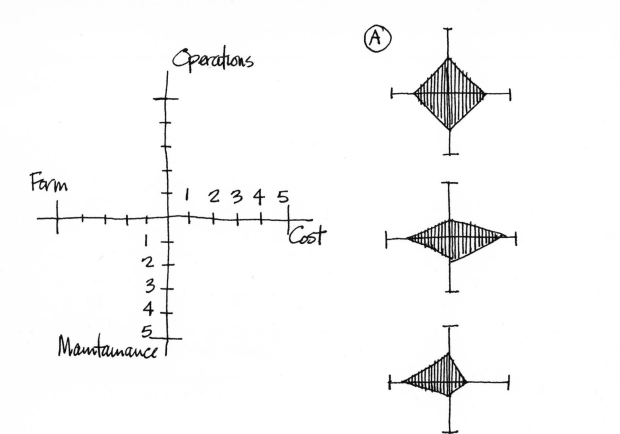

Operations

Farm

1 2 3 4 5

Cost

1
2
3
4
5

Maintainance

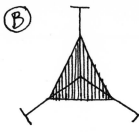

Ⓐ

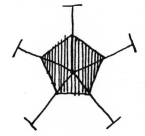

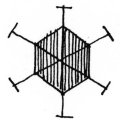

Ⓑ

73

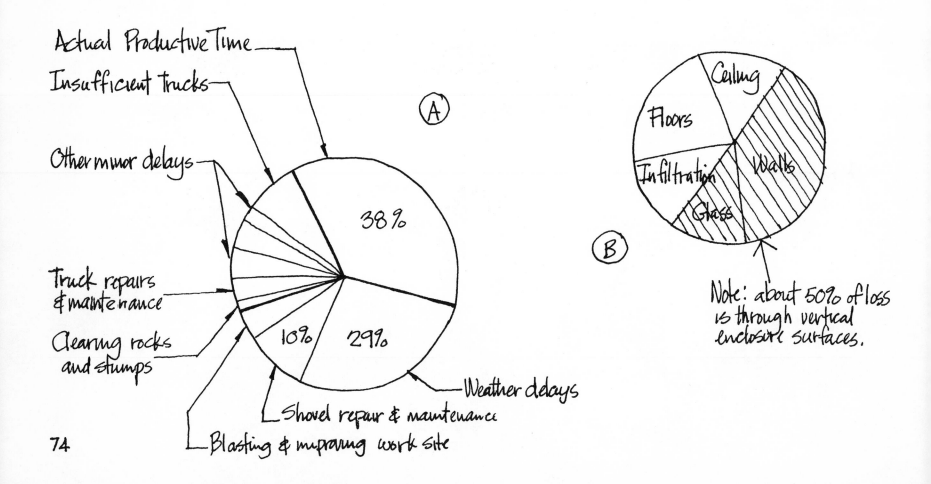

Actual Productive Time

Insufficient trucks

Other minor delays

Truck repairs & maintenance

Clearing rocks and stumps

(A)

38%

10%   29%

Weather delays

Shovel repair & maintenance

Blasting & improving work site

(B)

Ceiling

Floors

Infiltration   Walls

Glass

Note: about 50% of loss is through vertical enclosure surfaces.

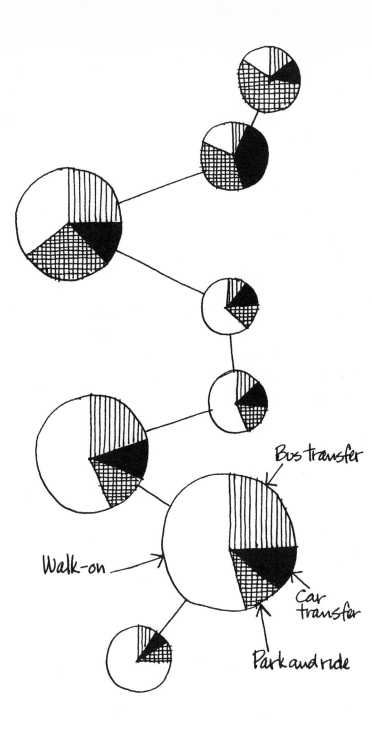

Bus transfer

Walk-on

Car
transfer

Park and ride

PIE DIAGRAMS
This is another type of radial diagram which
abstracts percentages or proportions. The examples
shown on the opposite page are an analysis of
productive time for a power shovel Ⓐ, and a
breakdown of normal heat loss through different
parts of a house enclosure Ⓑ.

TRANSPORTATION
The pie diagrams at the left characterize the
ridership originating at each station of a
proposed urban transit system. They indicate
passenger volume, type of transfer and sequence in
the transit system. Seeing the whole system in
such terms helps in setting design objectives
and priorities for each station.

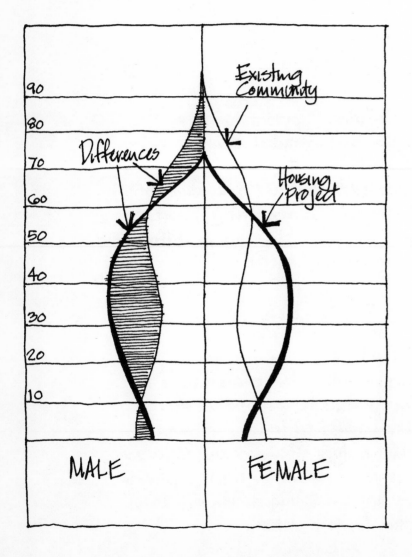

90
80
70 — Differences
60
50
40
30
20
10

MALE   FEMALE

Existing Community
Housing Project

POPULATION PYRAMID

A major concern of large in-city housing developments is their impact upon the existing community. A principal factor is the difference between the age distribution of the populations of the project and the community, as shown at left. The data for this diagram is easily derived from the census and project market analysis and may hold cues for the basic design approach.

CONSTRUCTION ACTIVITY

The diagrams on the right try to present an overview of the intensity of construction activity for a building project over a period of a few months. The bar chart shows volume of work for each trade and the building-plan diagrams show the general areas of construction activity. This combination of sketches might give subcontractors an added sense of the job.

FORM WORK   CONCRETE   CONCRETE   GLAZING
POURED-IN-PLACE PRE-CAST

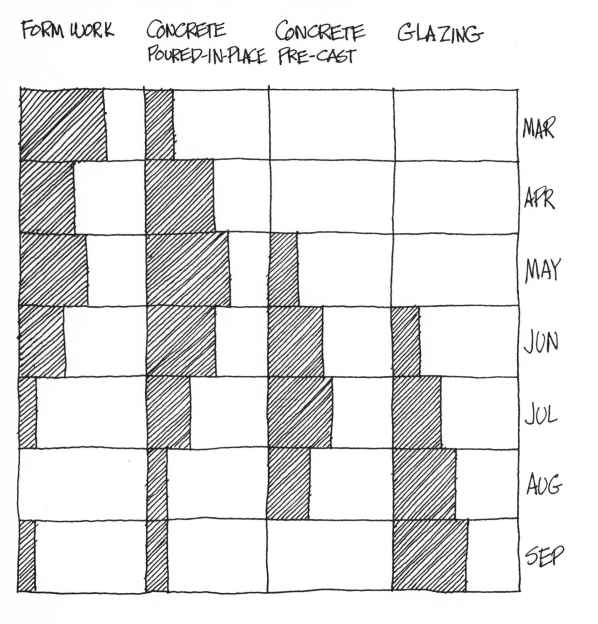

MAR
APR
MAY
JUN
JUL
AUG
SEP

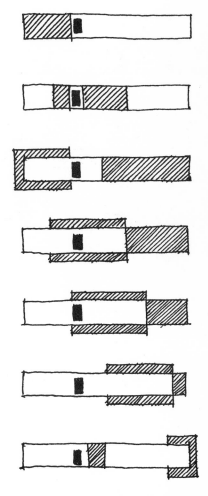

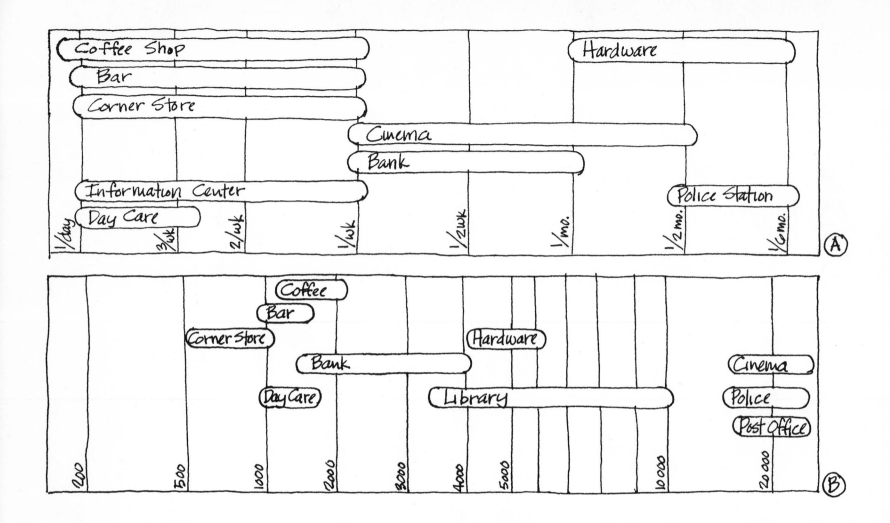

Presenting program information to designers in a way most useful to their task should be a concern of both researchers and designers. The bar charts on these pages show another approach to such a presentation. Frequency of use of a facility, whether existing or projected, is often considered by the designer. The advantage of the diagram Ⓐ is in showing the range of use by the spectrum of community inhabitants and the relationships between the use of several facilities. The diagram is also a convenient checklist of facilities required. The second diagram Ⓑ makes

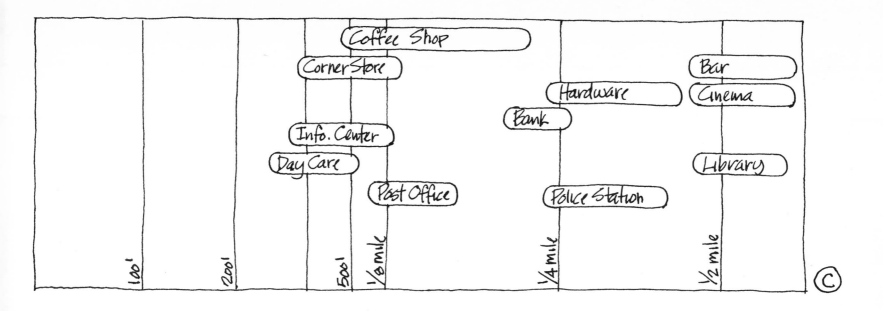

Bank : serves 1500 to 4000 people within a ¼ mile radius
with usage ranging from once per week to once per month.

Bar : serves 1000 to 1500 people within a ½ mile radius
with usage ranging from once per day to once per week.

a similar display of the number of people required to support different facilities. Here we can read the way certain population ranges make feasible the inclusion of both commercial and non-commercial facilities. The last diagram Ⓒ introduces a type of information often missed by designers.

Optimum walking distances from dwellings and travel time are important factors in trip decisions. By comparing charts Ⓑ and Ⓒ our assumptions about density of the proposed development can be checked against the optimum availability of amenities in several categories.

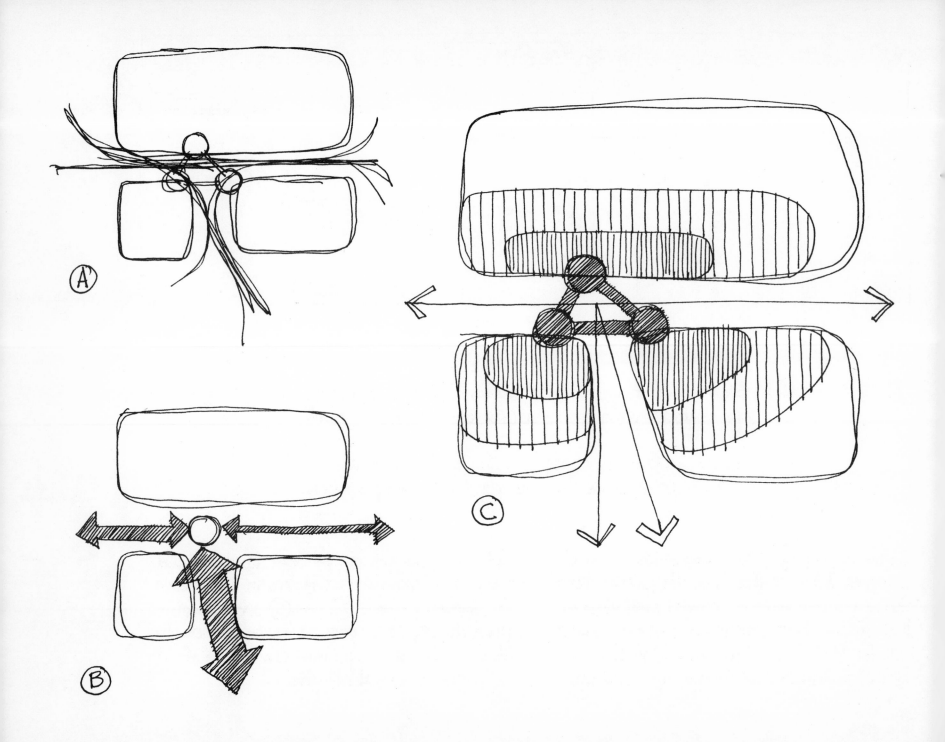

INTENSITY OF ACTIVITY

As designers we often try to get a "feel" for how a building might be used. The chicken-scratch diagram Ⓐ, a familiar type, shows various possible paths, indicating density or volume (roughly) by number of lines. Facilities programming has yielded intensity as measured by number of trips between points, which can be illustrated by thickness of line Ⓑ. Another approach Ⓒ depicts zones of activity with tones becoming darker as the degree of activity over time increases.

There are a couple of variations of the circulation diagram, which show traffic volume Ⓓ and Ⓔ. In these examples the designers were studying the central plan for a university campus. In the first diagram anticipated trip data are represented in a volume-flow diagram superimposed on the campus plan. In the other diagram traffic generators are added in the form of circles whose area corresponds to the relative amount of traffic originating at different points. It is possible to read the importance of both paths and destinations in the total campus circulation.

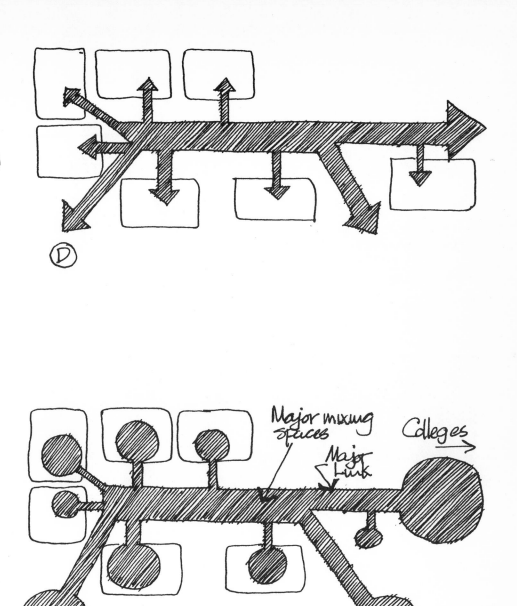

Ⓓ

Major mixing spaces

Major Link

Colleges

Parking

Commercial Center & Parking

Ⓔ

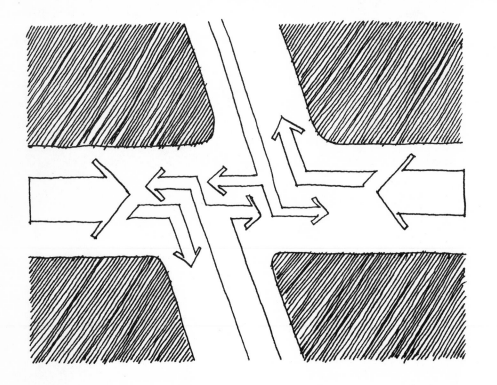

TRAFFIC VOLUMES

Whether dealing with auto traffic, on the left, or pedestrian traffic in a shopping center, on the opposite page, the summary of volumes and paths adds a level of understanding that can be quite valuable. The two-dimensional diagram is easy to construct and is based on an elementary type of traffic survey. The three-dimensional diagram takes a little longer to draw but is readable at once and therefore useful to the designer. The vertical scale in this diagram indicates the aggregate volume of pedestrian traffic passing a given point. The importance of different locations is based on the link between traffic and market volumes.

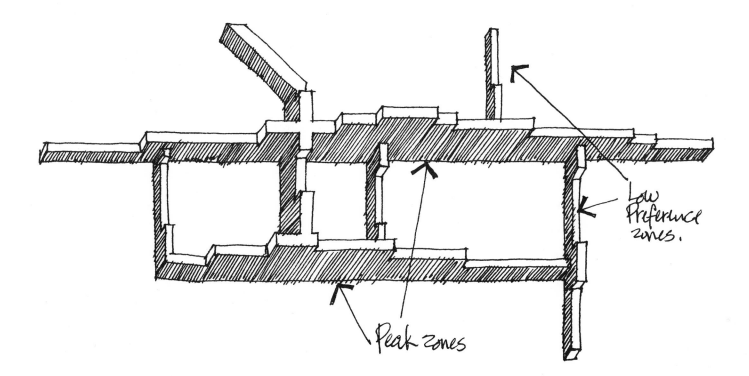

Low
Preference
zones.

Peak zones

83

CONDENSING CLIMATE INFORMATION

Energy conservation and climate control having emerged as major concerns, we now find it increasingly important to come to grips with how the many macro- and micro-climate factors act together as a major determinant of building form. Many site planners and landscape architects have led the way with their notation devices. As designers we are often concerned with an overall weather profile, which deals either with the change in a factor over time or with the relationship of factors. The significance of these diagrams, for the designer, is in the way he reads them. Issues covered in these diagrams are: temperature Ⓐ; precipitation Ⓑ; wind direction and speed Ⓒ; sun, air movement and acoustic conditions on the site Ⓓ.

84

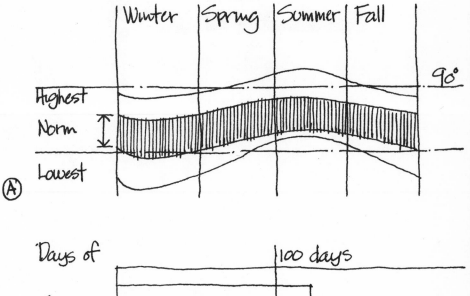

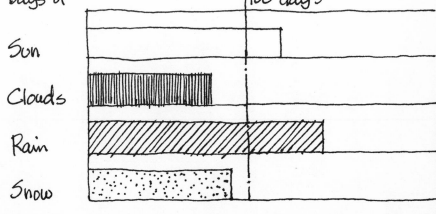

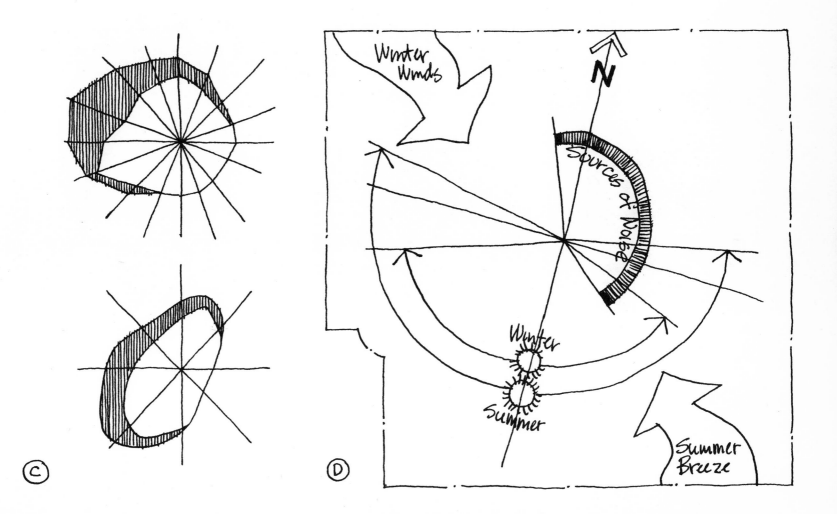

Winter
Winds

N

Sources of Noise

Winter

Summer

Summer
Breeze

©

Ⓓ

COST-BENEFIT ANALYSIS

When designing a house it is helpful to give the client a rough overall grasp of what he is getting for his money. The diagrams below relate the various spaces of the house directly to costs. Although the relationships of costs are more complex than this, given the interrelatedness of construction processes, these diagrams can still be useful.

The circle and square are two different ways to show the construction budget and its parts.

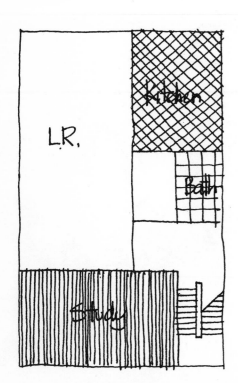

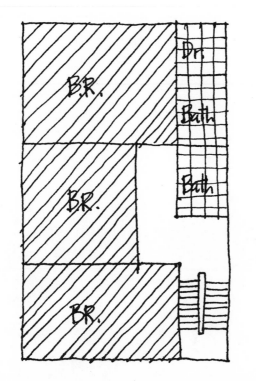

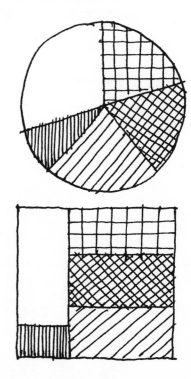

Trends

Most of the decisions we make as designers will affect our clients and others sometime in the future. The implicit premise of our work is that we are anticipating the interaction between people and the building. Clients are asking that we be more explicit about not only the future performance of a building but also its lifetime costs. Trend analysis with the aid of computers has become available to the building industry, but the effectiveness of such analysis will depend heavily upon our ability to interpret, to visualize its messages. This section shows some possible ways to use trend analysis. First two necessary warnings:

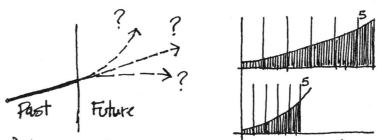

1) trend analysis is usually a guess at the future changes in a known <u>past</u> trend; 2) unconsciously we associate the slope of a curve with a rate of change. As shown above, the slope can be significantly altered simply by collapsing the time frame.

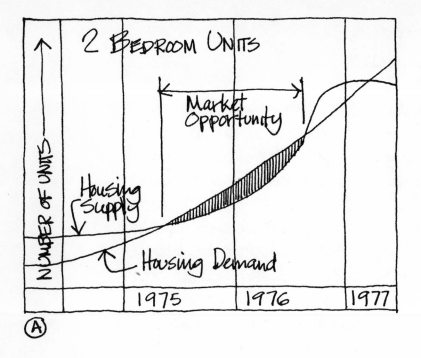

(A)

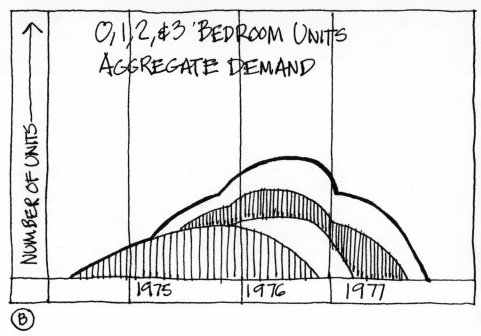

(B)

MARKET ANALYSIS

Simple supply and demand curves can be used to describe the size and duration of a market opportunity, in this case in the area of housing (A). By aggregating the opportunity zones (B) we can get a feel for the market in total or in parts over a period of time. This diagram does not represent detailed market analysis but is a generalized view of a market which helps us to understand some relationships.

AIRPORT PROGRAM ANALYSIS

The bulk of the users of an airport are in transit, on the move. The characteristics of pedestrian traffic are therefore important to many design decisions. Chart Ⓒ shows the fluctuation in volume of traffic, and chart Ⓓ shows the duration of stay at the airport for different total volumes of passengers.

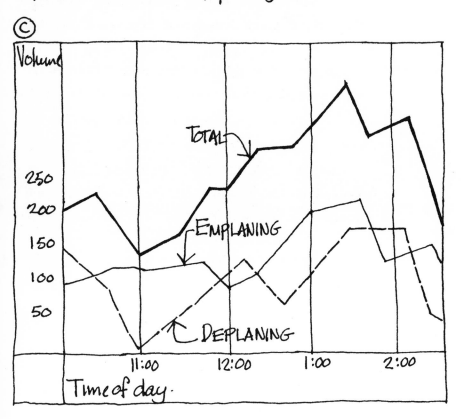

Ⓒ

Volume

250
200
150
100
50

TOTAL
EMPLANING
DEPLANING

11:00   12:00   1:00   2:00

Time of day.

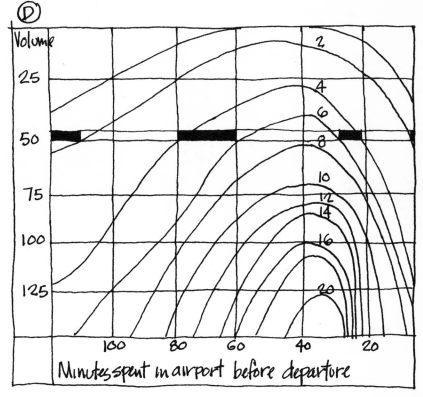

Ⓓ

Volume

25
50
75
100
125

2
4
6
8
10
12
14
16
20

100   80   60   40   20

Minutes spent in airport before departure

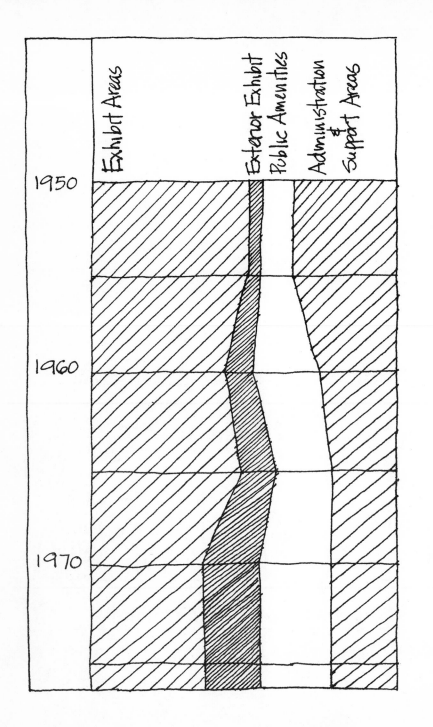

1950

1960

1970

Exhibit Areas

Exterior Exhibit
Public Amenities

Administration
& Support Areas

DESIGN TRENDS

The jokes about architects not visiting their old buildings really pointed at our disinterest in past problems and solutions. Recently many architects have been making use of past experience by transforming the pattern of past decisions into general guidelines for future work. This type of research is very helpful in making original estimates of space, land and equipment requirements as well as costs. The approach is to look at the way the same decision was made for several buildings rather than looking in detail at just the last building. The diagram at left looks at the change in area allocations for different functions in museums designed over the past several years. Changes in design could also be charted against geographical location or total size of the museum. The diagrams to the right indicate the ratio of different hotel facilities to the number of guest rooms. Besides making an approximation of facilities required for a certain size hotel, we can also see the relationship between minimum populations to support different facilities and trends in the hotel program as it increases in size. With some simple substitutions of factors these diagrams could be used for many other building types.

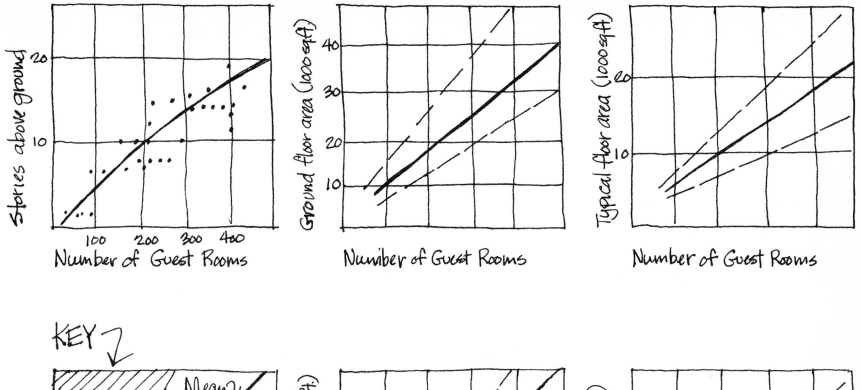

Stories above ground vs *Number of Guest Rooms* — 100, 200, 300, 400, 10, 20

Ground floor area (1000 sq ft) vs *Number of Guest Rooms* — 10, 20, 30, 40

Typical floor area (1000 sq ft) vs *Number of Guest Rooms* — 10, 20

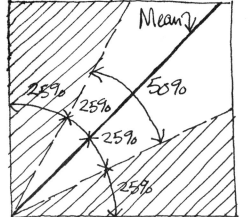

KEY

Mean

25% 25% 25% 50% 25% 25%

Percentages of Hotels tested as shown in each grid.

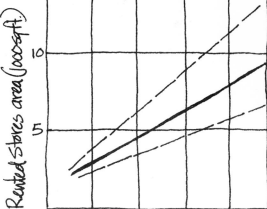

Lobby & Reception (1000 sq ft) vs *Number of Guest Rooms* — 1, 2, 3, 4, 5

Rented Stores area (1000 sq ft) vs *Number of Guest Rooms* — 5, 10

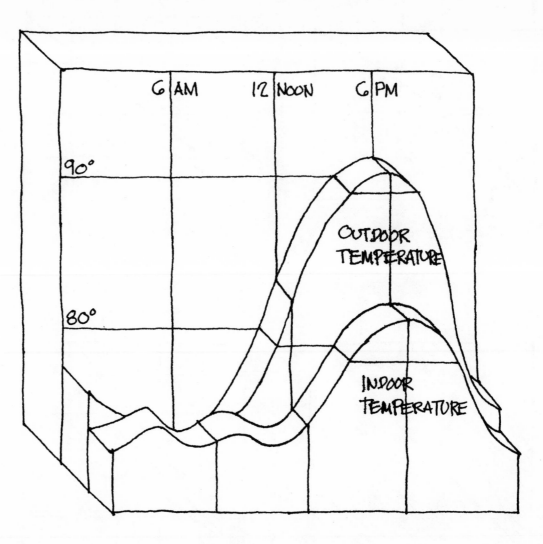

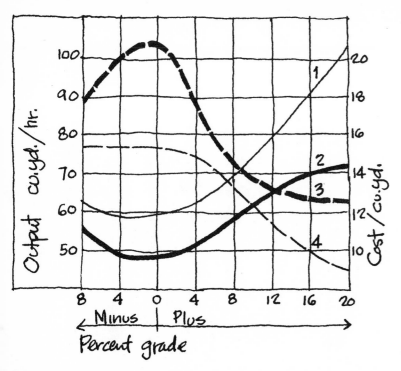

Key to curves:

1. cost for crawler-tractor unit
2. cost for wheel-tractor unit
3. output of wheel-tractor unit
4. output of crawler-tractor unit

CONTRASTING TRENDS

A distinct advantage of visual tools is their capacity to emphasize comparison or contrast. The diagram on the opposite page shows the relationship of indoor and outdoor temperature for one type of enclosure. A device often used by illustrators is the creation of volumes out of two curves, which tends for some reason to make them seem more real. They are simple to draw, being comparable to a 45-degree axinometric projection.

The chart at the immediate left compares the cost and performance of two types of equipment. It is really two charts in one: percent grade versus output and percent grade versus cost. Distinct identities for each curve are necessary in this situation.

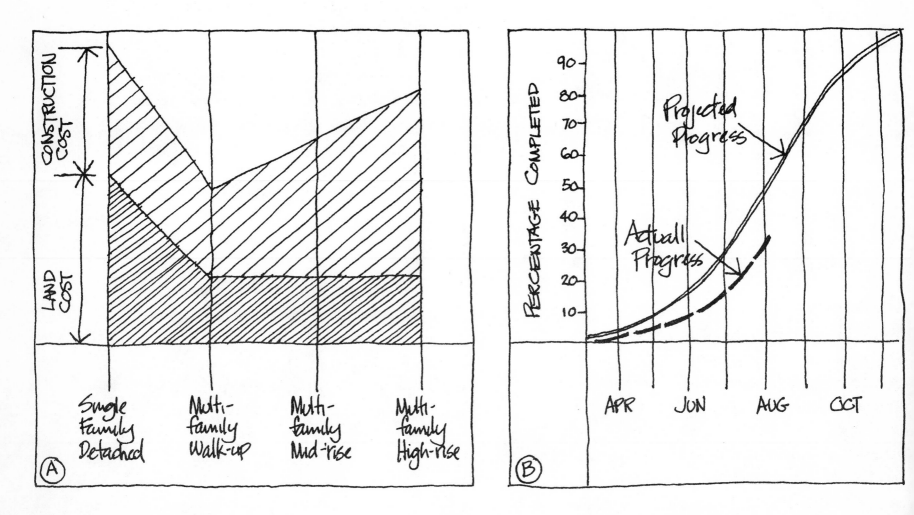

COST TRENDS

In addition to understanding the cost generators for a building type it may be important to see the fluctuation in costs that accompanies change in massing Ⓐ.

PROGRESS CHART

This device Ⓑ is a quick means for a contractor to sense how the job is going and make plans for completion on schedule.

STRUCTURAL DESIGN GUIDES

Here is an approach to summarizing design experience in order to make design assumptions on which to estimate costs without extensive calculations. It is possible to grasp in this way the relationships between beam depth and reinforcing bars required for a given span.

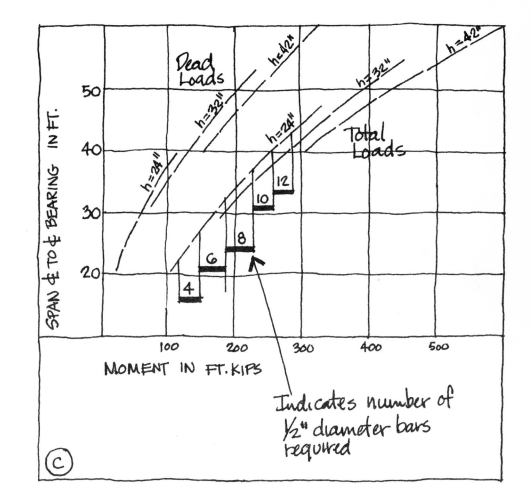

Indicates number of ½" diameter bars required

95

THREE - DIMENSIONAL CHARTS

Most charts relate two factors which have a constant rate of change; the range of acceptable relationships or solutions is found along a curve (A-1). Many fields have found use for a chart that relates two factors which do not have a constant rate of change. In this case the range of solutions is found within a plane and must be shown in three dimensions Ⓐ. A problem in using these diagrams for visual thinking is the difficulty of both drawing and reading the diagrams. One basic approach to resolving this problem is the use of contours Ⓑ. An advantage, of course, is the possibility of illustrating three-dimensional charts in two dimensions Ⓒ, similar to a site contour Ⓓ. In addition there are ways to render the three-dimensional qualities of the chart to enhance readability Ⓓ and Ⓔ. In the stepped model Ⓔ the coordinates are traced up the steps and an intermediate square footage estimated by triangulation.

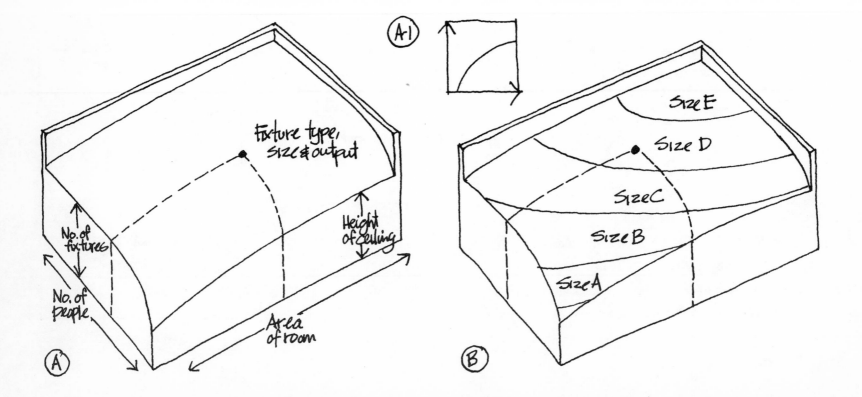

Ⓐ⁻¹

Ⓐ Fixture type, size & output — No. of fixtures — No. of people — Area of room — Height of ceiling

Ⓑ Size E — Size D — Size C — Size B — Size A

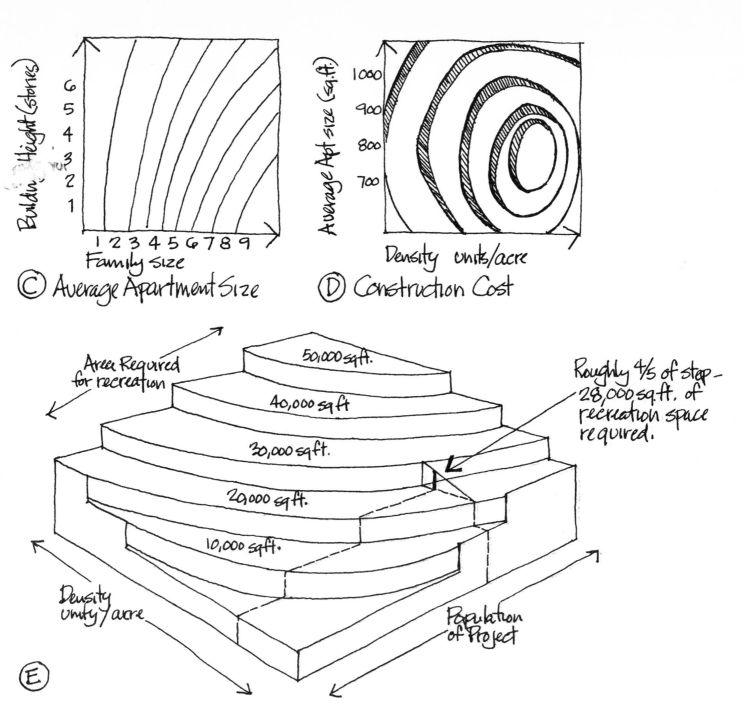

Building Height (stories)

6
5
4
3
2
1

1 2 3 4 5 6 7 8 9
Family size

© Average Apartment Size

Average Apt size (sq.ft.)

1000
900
800
700

Density units/acre

D Construction Cost

Area Required for recreation

50,000 sq.ft.

40,000 sq ft

30,000 sq.ft.

20,000 sq.ft.

10,000 sq ft.

Roughly 4/5 of step-
28,000 sq.ft. of
recreation space
required.

Density units/acre

Population of Project

E

97

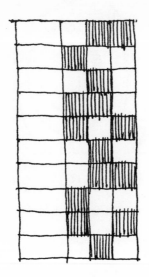

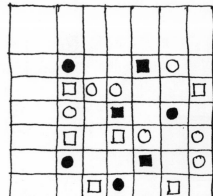

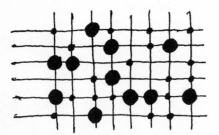

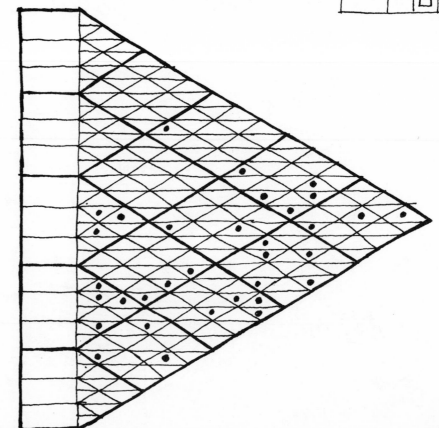

98

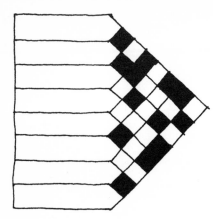

The Matrix

A matrix always sounded to me like something very mysterious and complex until I started using it. It is really a very simple tool and therefore useful for many things. A matrix is nothing more than a grid or a bunch of boxes whose use is up to us. Here is an example to get started:

	eggs	fish	turkey	cheese	potatoes
eggs				(a)	
fish					
turkey	(b)			(c)	
cheese					
potatoes	(d)		(e)	(f)	

The family cook is hard pressed to be inventive with today's rising food prices. Suppose we take a list of beef substitutes to see how many ways they can be combined: a) cheese omelet; b) turkey egg foo yang; c) turkey Tetrazzini; d) potato salad; e) turkey and mashed potatoes; f) potatoes au gratin. We have not invented any new dishes yet but the matrix has served to jog our memory by routinely matching food items with each other.

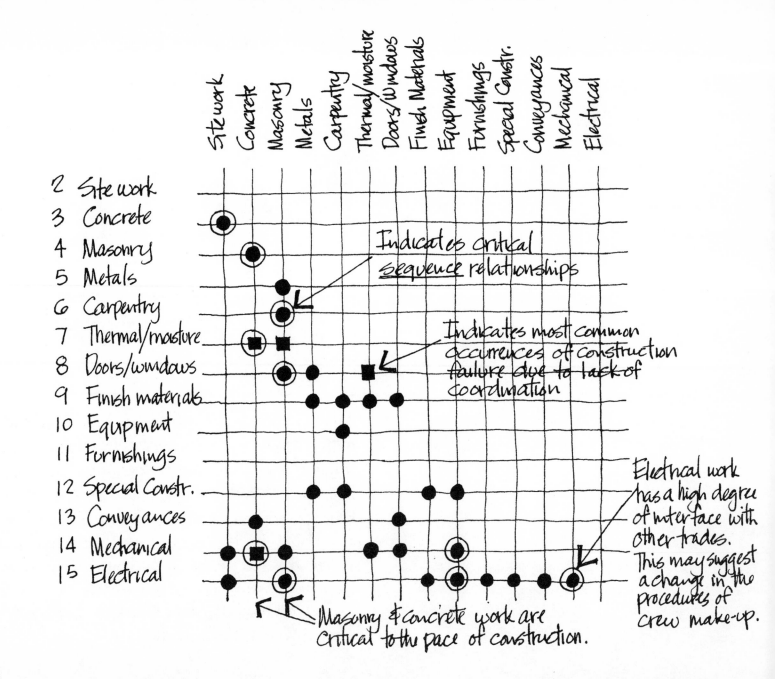

2 Site work
3 Concrete
4 Masonry
5 Metals
6 Carpentry
7 Thermal/moisture
8 Doors/windows
9 Finish materials
10 Equipment
11 Furnishings
12 Special Constr.
13 Conveyances
14 Mechanical
15 Electrical

Indicates critical <u>sequence</u> relationships

Indicates most common occurrences of construction failure due to lack of coordination

Electrical work has a high degree of interface with other trades. This may suggest a change in the procedures of crew make-up.

Masonry & concrete work are critical to the pace of construction.

This chapter is divided according to two basic uses of the matrix. First we will look at various ways in which a matrix can show a relationship or interface between things. Second we will be concerned with use of the matrix for comparative and evaluative decision making.

CONSTRUCTION MANAGEMENT

Good management anticipates problems and focuses on critical problems. An aid to both objectives might be a diagram of the way in which various construction trades interface or affect each other's work, opposite page. The manager is able to see potential sources of problems and areas where coordination is most needed. Management is also aided by each subcontractor seeing his role and the role of others in the total job. Each building may present a different profile so it could be equally important to record interface problems on the matrix during the period of construction.

BUILDING SYSTEM DESIGN

A similar use occurs when requesting proposals from industry for subsystems which will properly interface with the rest of the building to assure proper assembly.

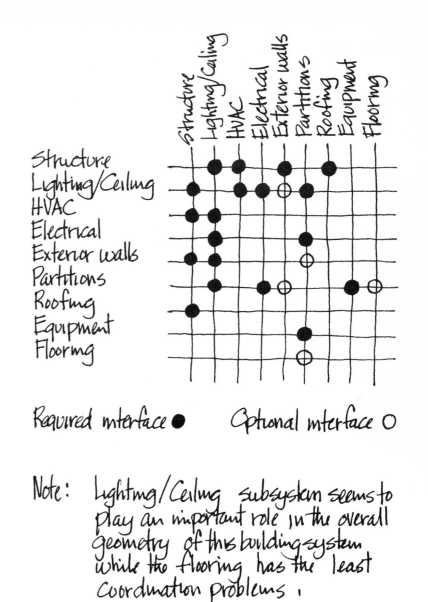

Required interface ●   Optional interface ○

Note: Lighting/Ceiling subsystem seems to play an important role in the overall geometry of this building system while the flooring has the least coordination problems.

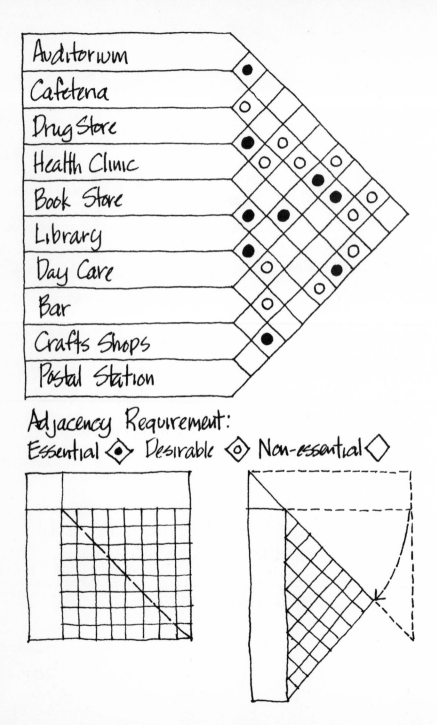

Auditorium
Cafeteria
Drug Store
Health Clinic
Book Store
Library
Day Care
Bar
Crafts Shops
Postal Station

Adjacency Requirement:
Essential ◆ Desirable ◉ Non-essential ◇

PROGRAM ANALYSIS

The matrix to the left may appear strange at first. It is really just half of a regular square matrix bent down so that the items being matched need only be listed once. (See below.) A similar diagram is used to list distances between major cities on some roadmaps. This matrix is being used to summarize the adjacency requirements for several activities in a community center. Note that different degrees of adjacency can be indicated.

GENERATING DESIGN IDEAS

"Morphological chart" is not simply new jargon. It is the correct name for the simple diagram drawn on the right. Its purpose is to open up our thinking to new ideas and areas of solutions to a given design problem. The approach is to break the problem down into its basic required functions. Then we list some alternative solutions for these functions. Next we explore different possible combinations of the subsolutions to uncover new solutions to the overall design problem.

REQUIRED FUNCTIONS	ILLUMINATION OF TABLES	ILLUMINATION OF CIRCULATION	EMPHASIS ON ENTERTAINMENT	INTIMATE ATMOSPHERE	CLEANING
SUB-SOLUTIONS 1	indirect lighting	indirect lighting	spot lights	colored lights	flood lights
2	table lamp	floor lights	dimmer adjustment	candles	fluorescent lights
3	candles	illuminated floor	concentration of fixtures	fireplace	incandescent downlights on dimmer
4			colored lights		

SPACE TYPES →					
ACTIVITIES ↓					
Offices	Single	Single or Double	Double		
Secretarial	2 typists	Reception			Typing Pool
Classroom				Seminar	Seminar
Laboratory		Single Private		Double	Group

SPACE/USE ASSOCIATIONS

This version of the matrix aids the design process by making an association between needs and solutions. Above we are concerned with the usefulness of different sets of space modules. At the right we explore the opportunity to maximize the use of existing community facilities as an alternative to new construction.

SPACE TYPES → ACTIVITIES ↓	Auditorium	Open Loft	Offices	Parking Ramp	Exterior Court
Commercial	(Movies Theatre)	Retail Stores	Professional offices	(Temporary market)	Outdoor Cafe
Government & Public	Town meeting Conventions	Community services, agencies	Administration, Services	(Celebrations)	Celebrations
Education	(Lectures, films, TV)	Laboratories		Exhibitions	
Recreation		Clubs, Crafts	Clubs, Crafts	Dances, Fairs	Resting area

It may be possible to use the same space at different times.

Might change the image of parking ramps

105

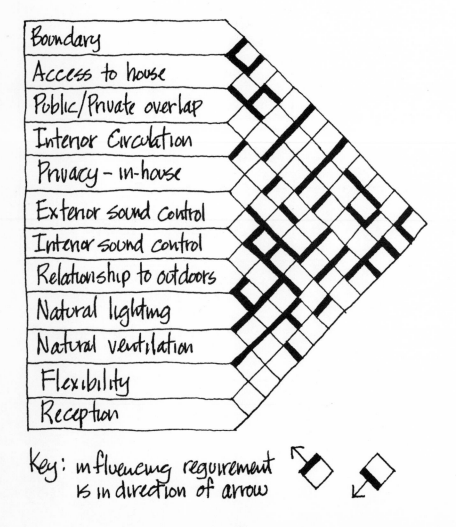

Boundary
Access to house
Public/Private overlap
Interior Circulation
Privacy - in-house
Exterior sound control
Interior sound control
Relationship to outdoors
Natural lighting
Natural ventilation
Flexibility
Reception

Key: influencing requirement
is in direction of arrow

DESIGN PROBLEM CLASSIFICATION

Another way to make decisions about the assignment of design tasks is to look closely at the program requirements. On the left requirements are matched and their effect on each other indicated. Requirements whose satisfaction influences the satisfaction of many other requirements should be studied. On the right we are concerned with the historical relationship between requirements and the spaces for a building type. As with the previous matrix we are really evaluating the design requirements. In addition we are using past experience to identify not only chronic problems but also design opportunities. As architects we should be as interested in realizing opportunities as we are in solving problems.

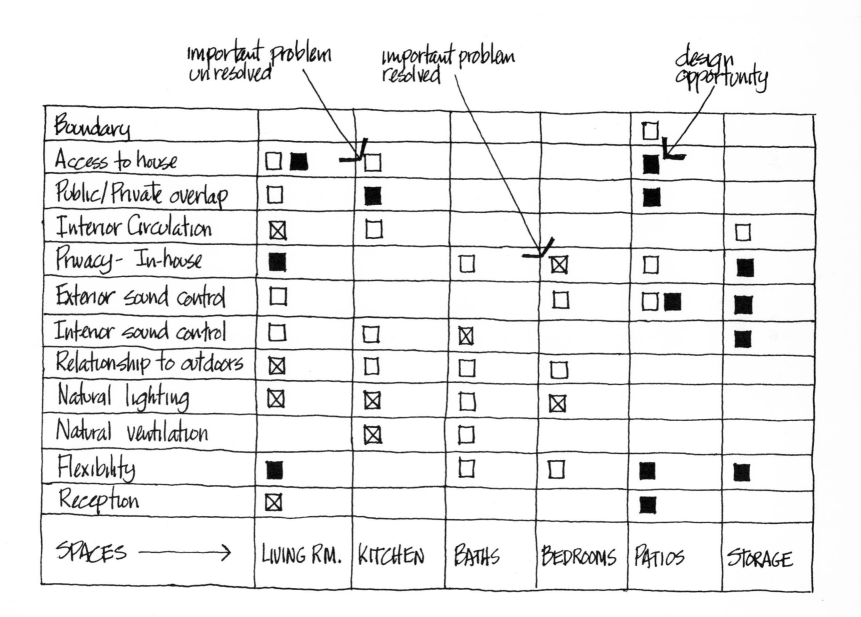

important problem
unresolved

important problem
resolved

design
opportunity

	LIVING RM.	KITCHEN	BATHS	BEDROOMS	PATIOS	STORAGE
Boundary					☐	
Access to house	☐ ■	☐			■	
Public/Private overlap	☐	■			■	
Interior Circulation	☒	☐				☐
Privacy - In-house	■		☐	☒	☐	■
Exterior sound control	☐			☐	☐ ■	■
Interior sound control	☐	☐	☒			■
Relationship to outdoors	☒	☐	☐	☐		
Natural lighting	☒	☒	☐	☒		
Natural ventilation		☒	☐			
Flexibility	■		☐	☐	■	■
Reception	☒				■	

SPACES ⟶

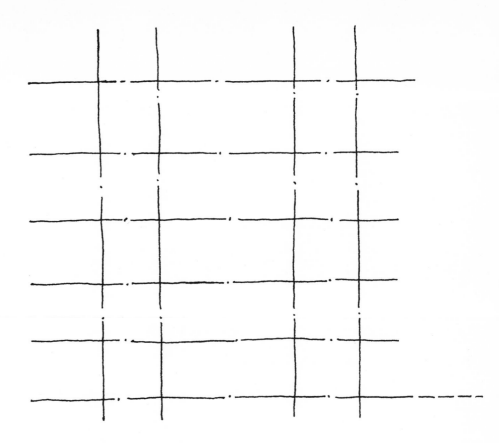

DESIGN GENERATION

A quite different use of the matrix is as a form of determinant to impose order upon a design at a very basic level. The grids shown here, structure above and zoning at the immediate right, are only two of the possible grids. These diagrams lend themselves to easy manipulation and an open range of interpretations, one of which is shown on the far right.

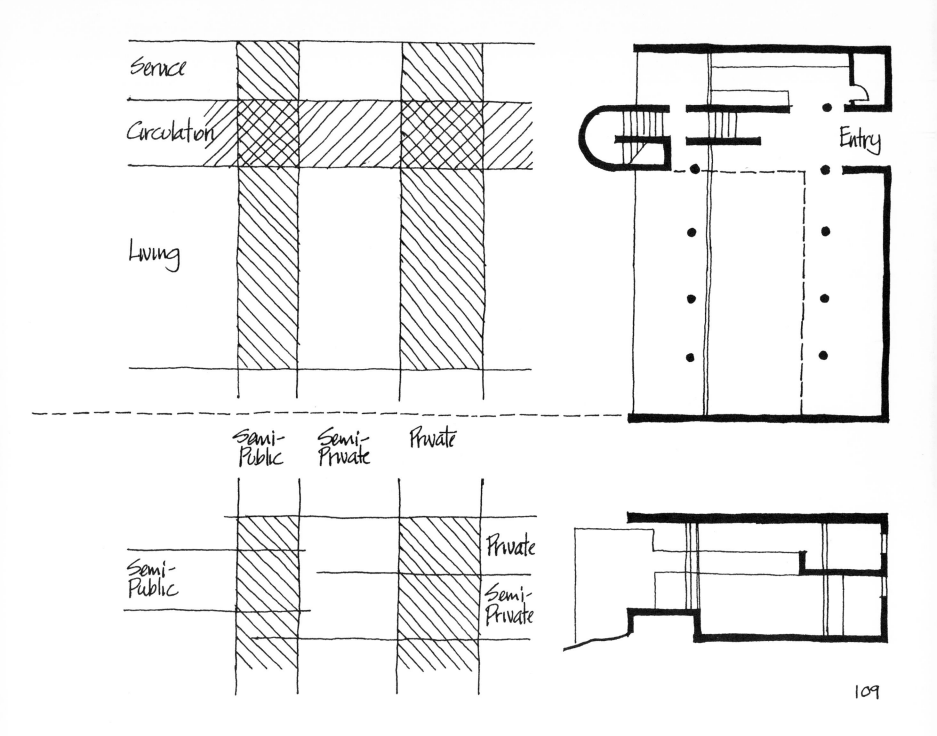

Service

Circulation

Living

Semi-
Public

Semi-
Public

Semi-
Private

Private

Private

Semi-
Private

Entry

109

	CONVENTIONAL 3 STORY HOUSE	BROWNSTONE	LARGE APT. HOUSE
ONE BEDROOM			
TWO BEDROOM			
THREE BEDROOM			
FOUR BEDROOM			

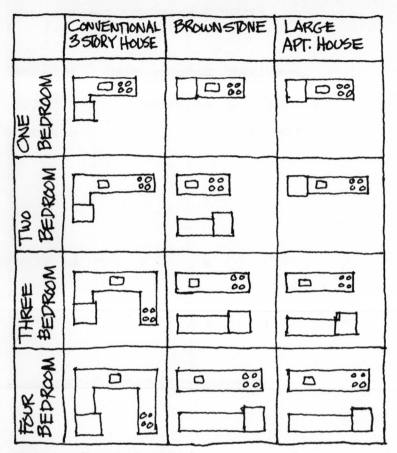

DESIGN OPTIONS

Using a format similar to the preceding pages
this matrix matches building types with apartment
types in a proposed rehabilitation project. The
purpose is to get a complete view of the range of
kitchen configurations. Using the matrix as a
reference tool the designer tries to increase the
number of options available while reducing the
number of components required.

Comparisons

Many people today and particularly builders and architects are deeply concerned about steps we can take to achieve a quality environment. Two approaches are being promoted: 1. establishing a set of qualitative criteria which are as authoritative as quantitative criteria for making building decisions. This has been a difficult task and probably at best is a long-range strategy that may require a a change in the values of society. 2. improving the impact and effectiveness of individual expert opinion on qualitative issues. This may be a more promising short-range strategy. Of the visual tools available, I find the matrix the most useful to the second approach. The matrix can make visible a pattern of opinions facilitating a comprehensive qualitative analysis. The task of evaluation, as shown in the following pages, requires 1) an explicit statement of criteria to be used for evaluation; 2) determination of the order of importance of the criteria; 3) assessment of the alternatives with respect to each separate criterion; 4) Comparison of patterns of response by alternatives to the full range of criteria.

HOUSING EVALUATION

Buying a house is a good example of a problem we often confront, namely, a discrepancy between what we want and what is available. It is a good example because we all consider it more seriously than practically any other purchase we make during our lives. The first step, one often overlooked, is to make the hard decisions about the priority of the things we want. (Would we give up the patio before the extra bedroom?) The small diagram below shows a technique for setting priorities. Requirements are matched on the matrix; a dot is made at each interaction box where a requirement from the side column supercedes a requirement from the top column. The more important requirements from the side column will have the most dots after them.

In the matrix at the right the list of requirements is in a descending order of priority and includes both house features and facilities we want within walking distance. Comparing the three houses on this matrix, we can immediately see the following:

a) Although the first house does not have some of the most desired facilities, it provides more overall facilities than the other two houses.

b) The second house provides the greatest number of high-priority items. This is readily apparent because of the density of dots toward the top of the matrix.

c) Many of the items missing in the third house could be compensated for by the purchase of a new car. This would not yield the same result in the case of the first house.

	large yard	new bath	cleaners	quiet street	
large yard			●		1
new bath	●		●	●	3 ← Highest priority
cleaners					0
quiet street	●		●		2

112

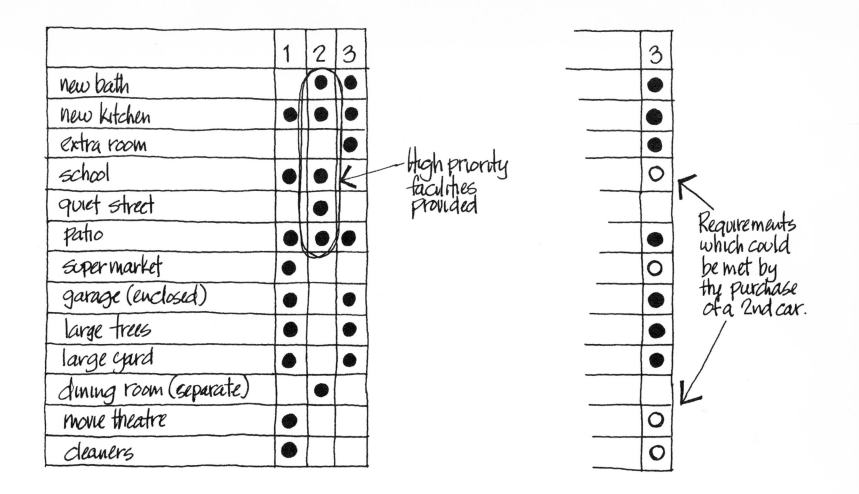

	1	2	3
new bath		●	●
new kitchen	●	●	●
extra room			●
school	●	●	
quiet street		●	
patio	●	●	●
supermarket	●		
garage (enclosed)	●		●
large trees	●		●
large yard	●		●
dining room (separate)		●	
movie theatre	●		
cleaners	●		

High priority facilities provided

	3
	●
	●
	●
	○
	●
	○
	●
	●
	●
	○
	○

Requirements which could be met by the purchase of a 2nd car.

A

B

C

	FORM				FUNCTION						SITE			ECONOMY				
	Memorable image	Scale / context	Scale / interior	Expression of functions	Communal spaces	Circulation	Security & Control	Indoor-Outdoor relationship	Access	Orientation	Utility of exterior spaces	Relation to other buildings	Relation to street	Economics of construction	Site preparation	Heat conservation	Operation maintenance	
A	⊙	O		⊙	O		⊙					O	O	O	O		⊙	
B			O	O		O		⊙	⊙	O	⊙	O	O					
C	⊙	⊙	O	O	⊙	O	O	O	O	O	⊙	O	O	⊙	O	⊙		

O = average quality   ⊙ = superior quality

DESIGN EVALUATION

Given three approaches to the basic plan for a small kindergarten complex, we try to get an overall comparison with respect to the issues of form, function, site and economy. Evaluation criteria or measures are listed for each issue. (The list above is representative rather than comprehensive.)

The three plans are compared with each other for each criterion, identifying best, worst and average. Evaluations could be given numerical values from one to ten and each scheme given a total score. As it is we can get a picture of the strengths and weaknesses of each scheme and make a decision about which should be developed further.

PRODUCT EVALUATION

A similar type of matrix can be used for the evaluation of products such as ceiling and lighting subsystems in the hypothetical examples below. The criteria here are stated so that only a yes or no statement can be made about the performance of each subsystem. The advantage of a matrix is that we can get a general sense of the acceptability of different products using an extensive range of requirements. The point I want to emphasize again is the potential of visual techniques to condense large amounts of information to a manageable size. The use of the diagram below is found more in the general pattern rather than in any single criterion.

	Compatable with structural	Compatable with HVAC	Compatable with partitions	has integrated incandescent lighting	incorporates power outlets	adjustable without special equipment	adequate acoustical barrier	uses standard acoustical tile	sufficient range of surfaces	adequate illumination at 16'-0" high	indirect lighting option	diffuser panel option	available nationally	less than 6 month delivery	less than 3 month delivery	local assembly contractors	technical consultants provided
5'-0" x 5'-0" Coffer Ceiling	■			■		■				■			■			■	
2'-0" one way grid		■	■	■		■					■		■			■	
2'-6" x 5'-0" two way grid	■		■	■		■					■			■		■	

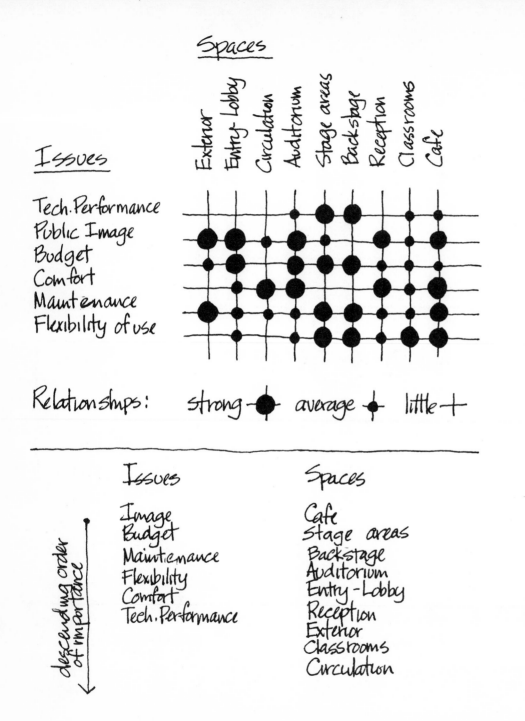

Spaces

Issues

Tech. Performance
Public Image
Budget
Comfort
Maintenance
Flexibility of use

Relationships: strong ● average · little +

Issues

Image
Budget
Maintenance
Flexibility
Comfort
Tech. Performance

descending order of importance

Spaces

Cafe
Stage areas
Backstage
Auditorium
Entry-Lobby
Reception
Exterior
Classrooms
Circulation

SETTING DESIGN PRIORITIES

Design priorities are often set on an intuitive basis; however, it may be important to test our intuitive judgements from time to time in order to gain fresh insights about a building type. The matrix, in this case, starts as a shopping list of design issues and major building spaces. At each point of relationship in the matrix we ask the question about importance of this issue in this space: strong, average or little. The degree of importance is indicated by the size of the circle. (I prefer circles of different size diameters over the use of numerical values simply because the matrix can be "read" visually.) At this point the exercise will present some priorities, which can be listed for both spaces and issues. But there are some further observations that can be made.

When the matrix is redrawn, on the opposite page, to reflect the descending orders of importance of spaces and issues, certain patterns emerge. By noting these patterns additional insights into the problem may be gained. Perhaps there is a basis for a different zoning of the building or a re-examination of one of the issues.

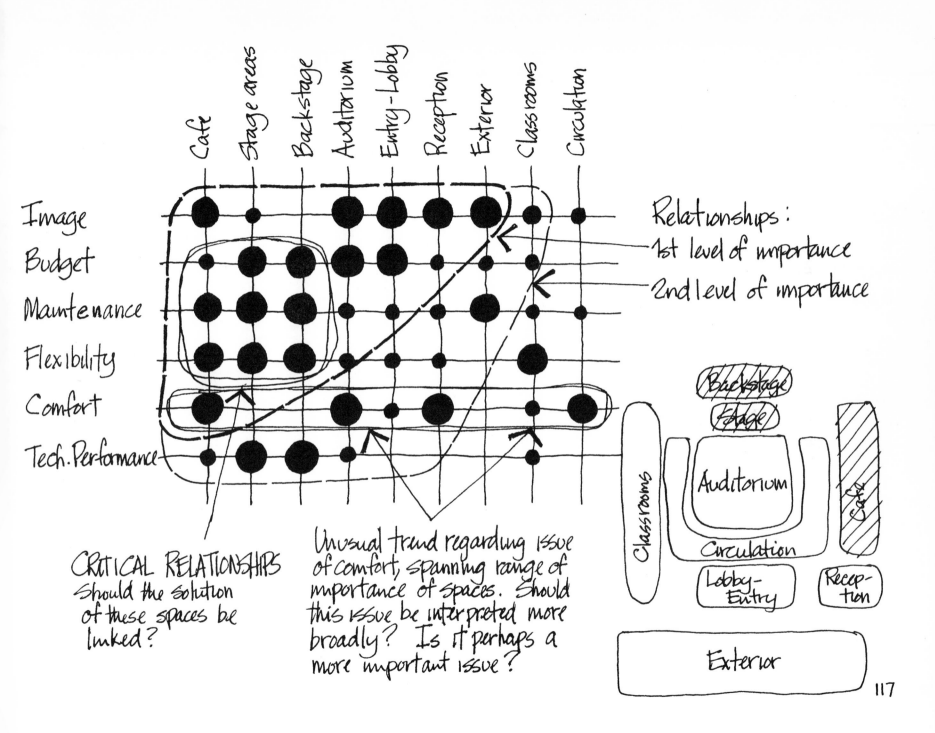

Cafe · Stage areas · Backstage · Auditorium · Entry-Lobby · Reception · Exterior · Classrooms · Circulation

Image
Budget
Maintenance
Flexibility
Comfort
Tech. Performance

Relationships:
1st level of importance
2nd level of importance

CRITICAL RELATIONSHIPS
Should the solution
of these spaces be
linked?

Unusual trend regarding issue
of comfort, spanning range of
importance of spaces. Should
this issue be interpreted more
broadly? Is it perhaps a
more important issue?

Backstage
Stage
Classrooms
Auditorium
Cafe
Circulation
Lobby-Entry
Reception
Exterior

117

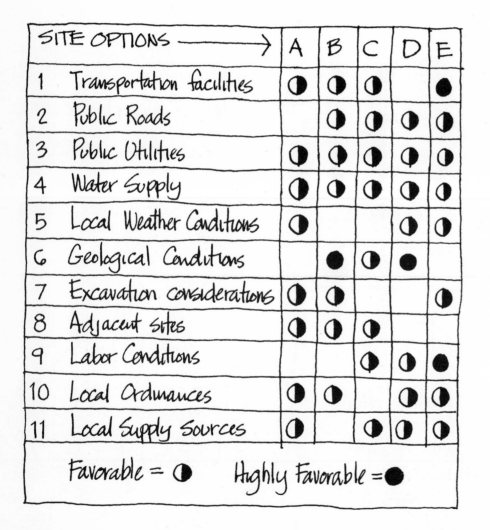

SITE OPTIONS →	A	B	C	D	E
1 Transportation facilities	◑	◑	◑		●
2 Public Roads		◑	◑	◑	◑
3 Public Utilities	◑	◑	◑	◑	◑
4 Water Supply	◑	◑	◑	◑	◑
5 Local Weather Conditions	◑			◑	◑
6 Geological Conditions		●	◑	●	
7 Excavation considerations	◑	◑			◑
8 Adjacent sites	◑	◑	◑		
9 Labor Conditions			◑	◑	●
10 Local Ordinances	◑	◑		◑	◑
11 Local Supply Sources	◑			◑	◑	◑

Favorable = ◑   Highly Favorable = ●

SITE EVALUATION
When included in the design team the builder can lend another dimension to many decisions through contracting criteria. At left several sites can be compared in light of cost-related conditions.

EVALUATION PROFILES
When we are presented with several options it can be useful to get a capsule or summary evaluation of each for comparison purposes. In the example at right nine building systems have their subsystems matched against important characteristics. By reducing the number of characteristics and the elements evaluated for each building system, the profile is readable. Likewise when the profiles are arranged in a set it is possible to make several general comparisons and conclusions.

SUB SYSTEMS→ CHARACTERISTICS ↓	Structural	HVAC	Lighting/Ceiling	Exterior Enclosure
Performance				
Compatibility				
Cost				
Maintainability				

O = Acceptable

● = Better than Average

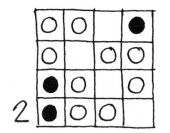

1

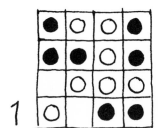

2

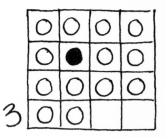

3

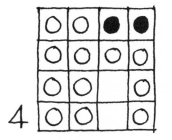

4

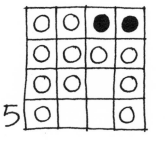

5

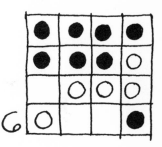

6

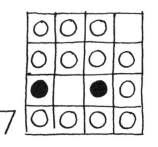

7

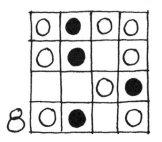

8

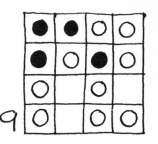

9

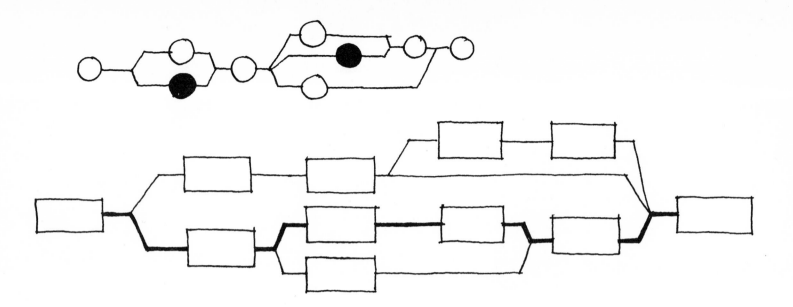

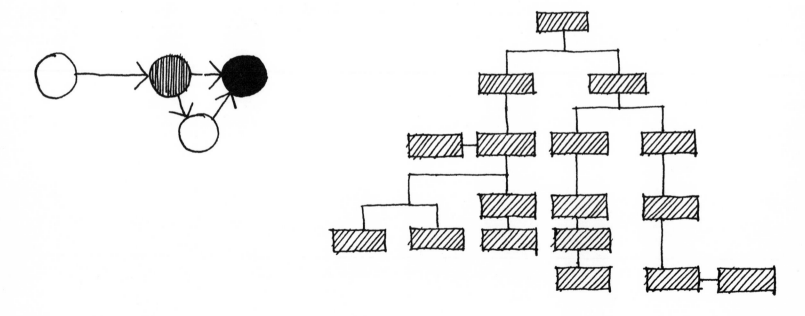

120

Networks

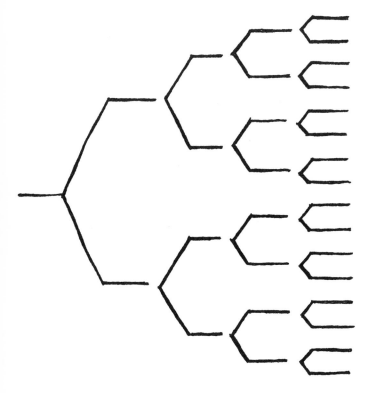

Surely the most popular device to come out of management science has been the network, be it critical path, PERT, or decision tree. While networks are more and more in use they seem to be less and less understood or really useful. My purpose here is to convert networks from their role as wallpaper and iconography to principles that are a part of everyday thinking. The first consideration in this regard is simplicity. Many networks read like a plate of spaghetti, chaotic and uncommunicative. This chapter will deal with the reduction of the quantity of information in a network, but the principles of visual communication can also be applied to make complex networks quickly understandable. The following pages show the origin and basic use of the network. Next the more extended type of network is reviewed and finally we will explore the potential uses of a special form of network, the decision tree.

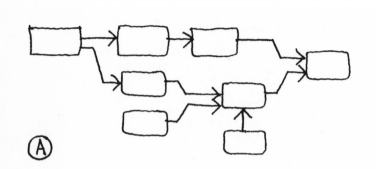

(A)

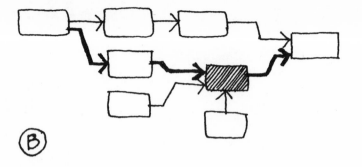

(B)

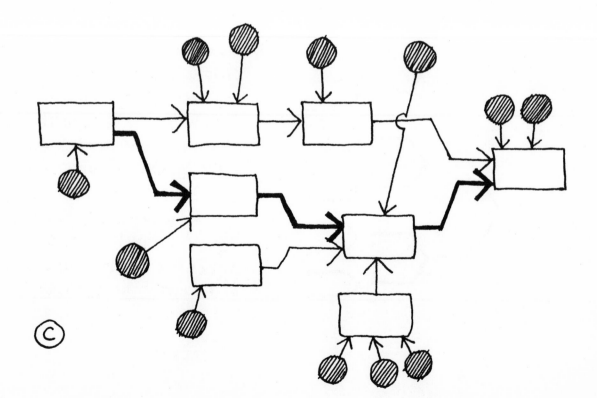

(C)

122

EVOLUTION OF NETWORKS

Although management planners often use bobble diagrams similar to those used by architects to show organization structure, a more distinctive language has been developed in order to model processes. The basic planning task for an assembly line is to get the right parts to the right point at the right time with minimum use of space and minimum conflict between operations. In order to solve this problem the assembly process is modeled in the form of a linear network Ⓐ. This shows the sequence of the basic operations. Next, critical operations and sequences are identified in order to determine labor and equipment requirements Ⓑ. In the final step the entry of materials or parts into the process is indicated Ⓒ.

Networks were also developed within the building industry. Their evolution from bar chart scheduling is show on the right.

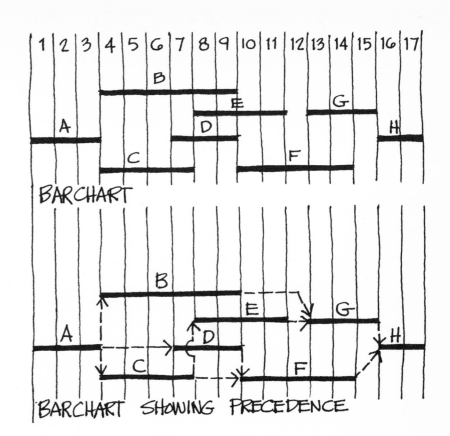

BARCHART

BARCHART SHOWING PRECEDENCE

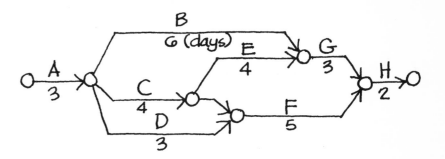

NETWORK

123

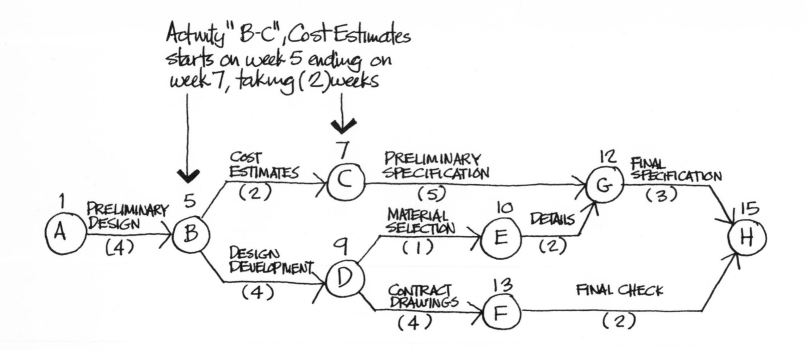

Activity "B-C", Cost Estimates starts on week 5 ending on week 7, taking (2) weeks

DIFFERENT TYPES OF NETWORKS

The language of networks consists principally of lines and bubbles. There are two basic types of networks: One, above, in which lines represent activities; and the other, opposite page, in which tasks or activities are represented by bubbles.

I prefer the latter type of network because it is easy to identify tasks and add information (see bottom of opposite page). The essential notation in both networks for each task is its duration, time starting and time finishing and, of course, the title or description.

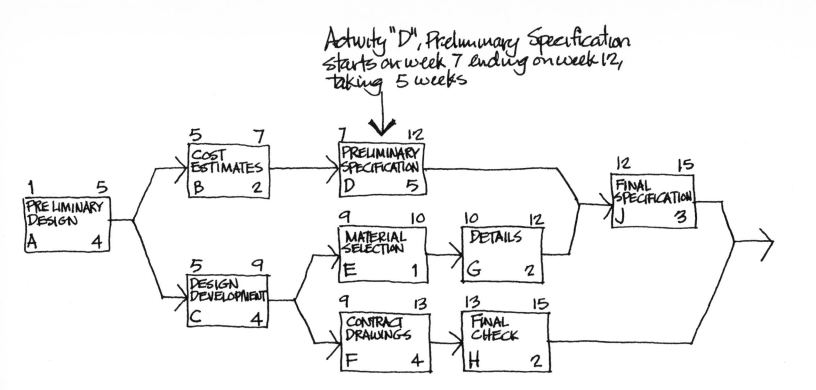

Actvity "D", Preliminary Specification starts on week 7 ending on week 12, taking 5 weeks

	5	7
	COST ESTIMATES	
	B	2

	7	12
	PRELIMINARY SPECIFICATION	
	D	5

	12	15
	FINAL SPECIFICATION	
	J	3

	1	5
	PRELIMINARY DESIGN	
	A	4

	9	10
	MATERIAL SELECTION	
	E	1

	10	12
	DETAILS	
	G	2

	5	9
	DESIGN DEVELOPMENT	
	C	4

	9	13
	CONTRACT DRAWINGS	
	F	4

	13	15
	FINAL CHECK	
	H	2

STANDARD SHAPES

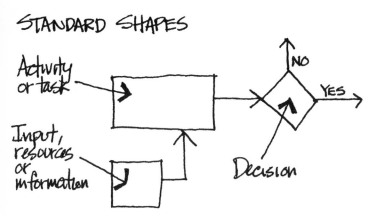

Activity or task

Input, resources or information

NO

YES

Decision

ALTERNATIVE FORMS

D	Preliminary Specification
	o General
	o Architectural
	o Mechanical
	o Electrical

D	7	Preliminary Specification	5	12

125

SIMPLIFYING NETWORKS

The difficulty with many networks available in reference works, such as the one below, is that too much information is provided. We can't retain it and we don't know what is critical and what is of secondary importance.

The first step toward simplification and clarification is the segregation of elements (B). This is followed by the delineation of the basic functions and essential connections (C). This diagram is now ready for conceptual design exploration and manipulation (D).

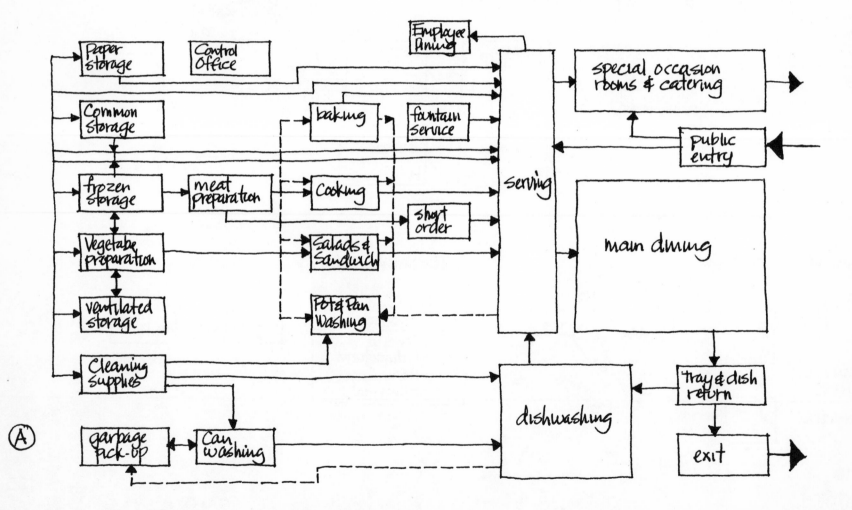

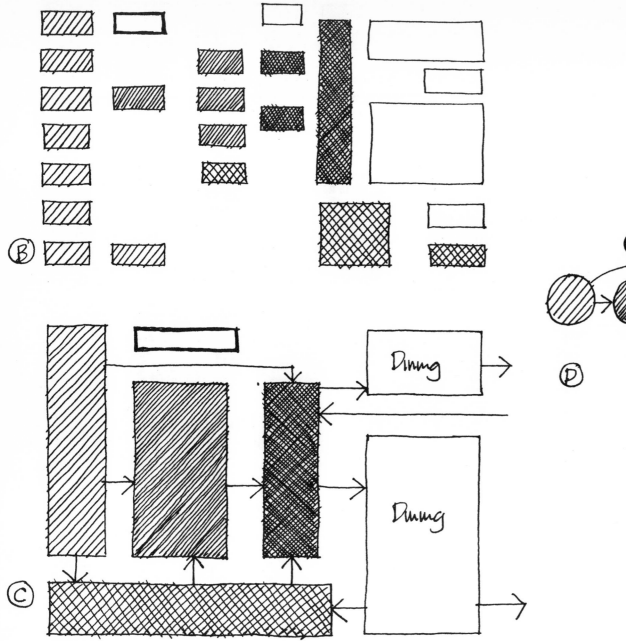

Ⓑ

Ⓒ

Dining

Dining

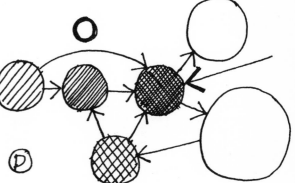

Ⓓ

127

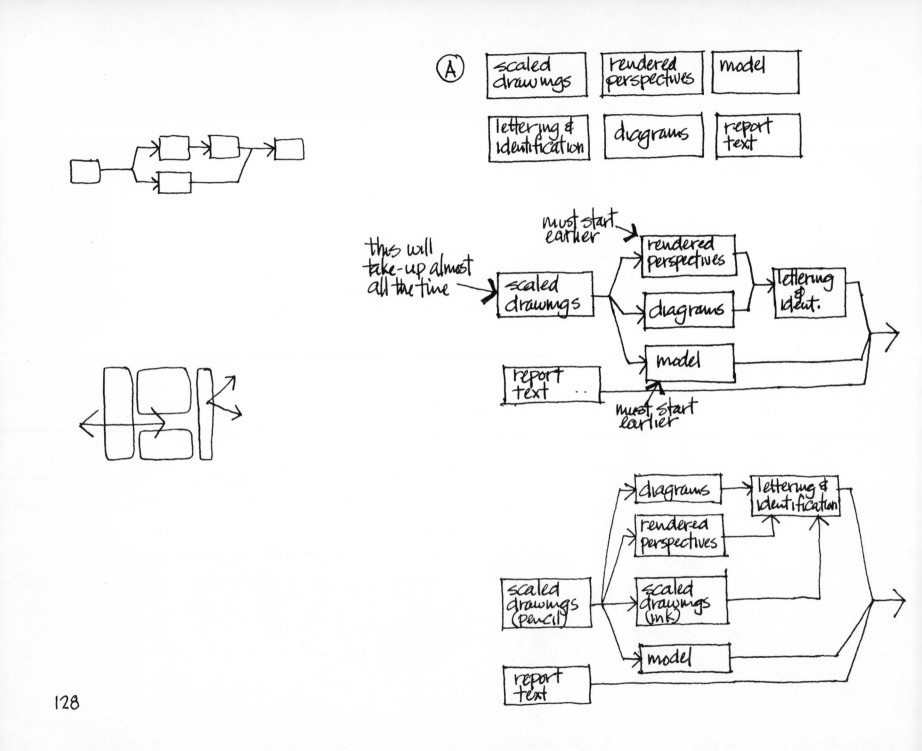

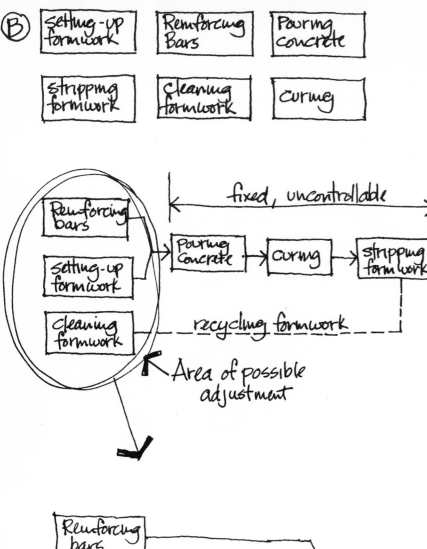

Ⓑ

| setting-up formwork | Reinforcing Bars | Pouring concrete |
| stripping formwork | cleaning formwork | curing |

fixed, uncontrollable

Reinforcing bars / setting-up formwork / cleaning formwork → Pouring Concrete → curing → stripping formwork

recycling formwork

Area of possible adjustment

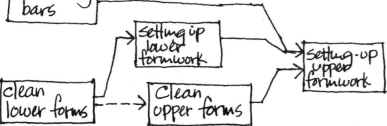

Reinforcing bars → setting up lower formwork → setting-up upper formwork

clean lower forms → Clean upper forms → setting-up upper formwork

Five-Box Diagram

A five-box diagram is a network that does not necessarily have five boxes, but it is an attempt to describe any process in the smallest number of boxes possible, in the neighborhood of five. This approach provides the kind of abstraction similar to that found in design parti sketches. This type of network can be useful because it will not allow us to become engrossed in network graphics, and it forces us to be specific about basic decisions.

In the first example Ⓐ a decision is being made about the allocation of people on a project presentation that has a severe time constraint. The second example Ⓑ looks at a recurring cycle of operations related to poured-in-place concrete work. In both examples the basic tasks are identified. Then they are placed in a network to show their ideal sequence. This network is reviewed with respect to particular constraints on time or manpower. The final revised network shows the adjustments necessary to be successful within the constraints.

129

Report on Environmental Analysis

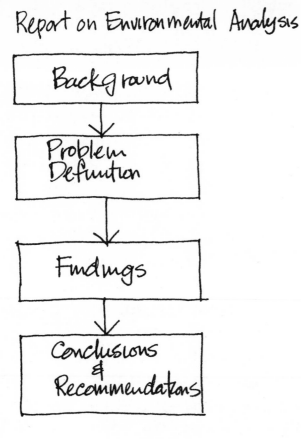

Design Presentation

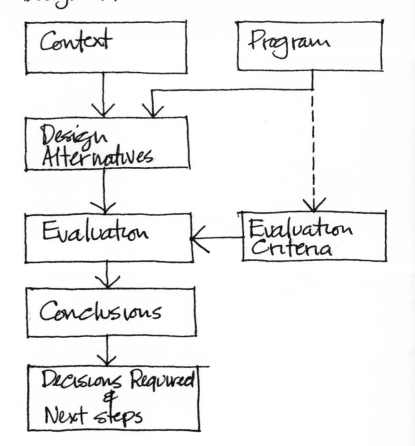

REPORTS AND PRESENTATIONS
Verbal or written presentations to clients are linear in nature; that is, ideas must be presented in a sequence; they cannot be presented simultaneosly as in a drawing.

The order in which ideas are conveyed is important to our effectiveness. Diagrams such as those above are helpful in planning presentations and in giving the client a synopsis at the introduction of a presentation or report.

Expanded Network

Assuming an understanding of simple networks, let's tackle some more complex uses of networks. A reminder, however: All of the following networks can be boiled down to the five-box diagram. We should always be in the habit of reading networks for their simplest form. In some cases I have shown miniature five-box diagrams next to the expanded networks. I have also incorporated, where possible, graphic devices to help in reading the basic structure of the expanded network. One such device is the grouping of boxes, another the emphasis of the border of the boxes, and a third device the marking of territories.

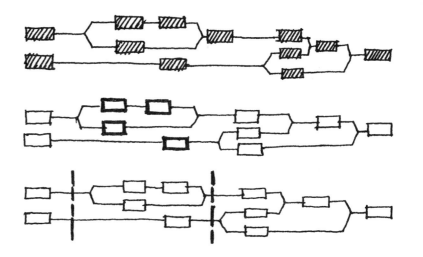

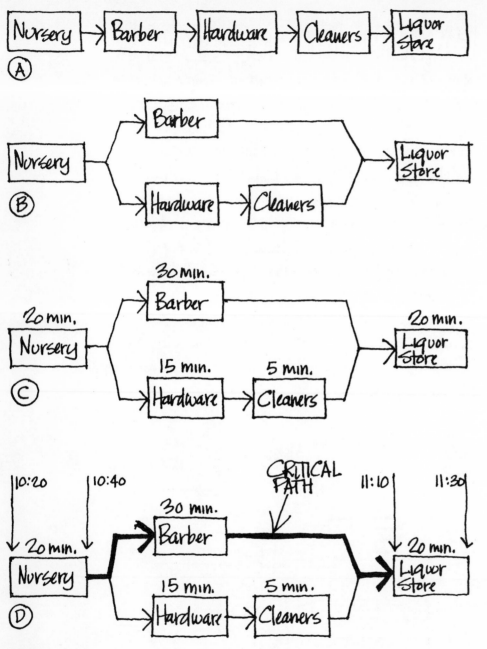

Nursery → Barber → Hardware → Cleaners → Liquor Store

(A)

Nursery → Barber, Hardware → Cleaners → Liquor Store

(B)

20 min. Nursery — 30 min. Barber — 20 min. Liquor Store; 15 min. Hardware → 5 min. Cleaners

(C)

10:20 10:40   CRITICAL PATH   11:10   11:30
20 min. Nursery — 30 min. Barber — 20 min. Liquor Store; 15 min. Hardware → 5 min. Cleaners

(D)

CRITICAL PATH

It is likely that most architects and builders are aware of the role of a critical path in a network. Simply put, it is the one sequence of operations whose total time determines how long it takes to complete the network. What some of us may not know is how to use the concept of critical path as an aid to thinking or problem-solving. Planning for a hypothetical Saturday morning trip will serve as an example of the role of critical path in everyday thinking. As with all networks I start with a checklist, and the first consideration is sequence (A). I determine the order for my errands by the location of stores (seeking the shortest and simplest round-trip) and other considerations, such as picking up the cold beer last. Next I try to save time on the total trip by looking at activities which can be conducted at the same time (B). These are called parallel activities. Now dead lines are familiar to us all. My wife reminds me that she needs the car by 11:30. I am in the middle of a leisurely breakfast, the first this week, so some time estimates run through my mind (C), (D).

Working back from 11:30 I find that 10:20 is the latest I can begin, because the visits to the nursery, barber and liquor store must be in sequence. This is the critical path because my son's stay at the barber is longer than the visits to the hardware store and cleaners. At 10:00 my wife eyes me impatiently and I mentally ask myself what could go wrong with my plan. The first thing that is apparent is the distinction between those activities over which I have greater or lesser control Ⓔ — nursery: I'm not sure how long it will take to find what I want; barber: time required depends on the number of people waiting. I also do a priority check. Which errands must absolutely be completed today Ⓔ? With this new perspective I make some adjustments in my schedule to improve my chances of a successful trip Ⓖ. The changed plan accomplishes the following: 1) removes the barbershop from the critical path of errands; 2) places the nursery at a point where it can be easily dropped (nursery is high risk, low priority); 3) reduces overall time required by ten minutes.

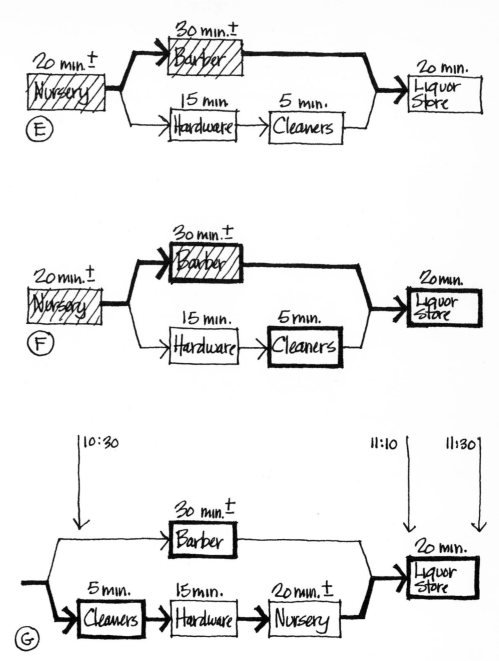

133

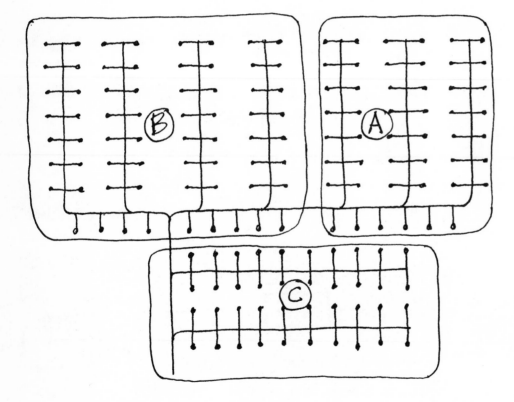

PLACING UTILITIES

Sometimes the scale of a project is of sufficient size so that it should be divided into sections. This allows a staggering of tasks, which can collapse the total time required in order to meet severe time or schedule constraints. This may also permit the reduction of crews and equipment and reduce idle time. The schedule can be easily worked out as on the opposite page.

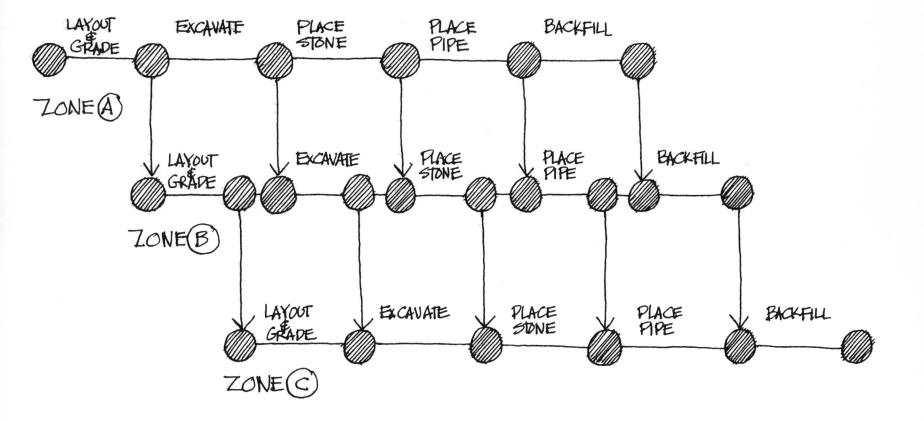

ZONE (A)

LAYOUT & GRADE · EXCAVATE · PLACE STONE · PLACE PIPE · BACKFILL

ZONE (B)

LAYOUT & GRADE · EXCAVATE · PLACE STONE · PLACE PIPE · BACKFILL

ZONE (C)

LAYOUT & GRADE · EXCAVATE · PLACE STONE · PLACE PIPE · BACKFILL

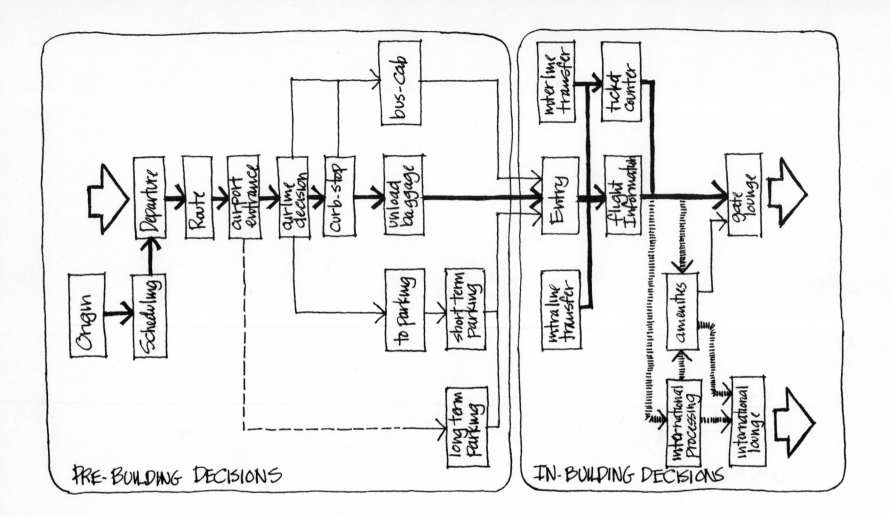

PRE-BUILDING DECISIONS

IN-BUILDING DECISIONS

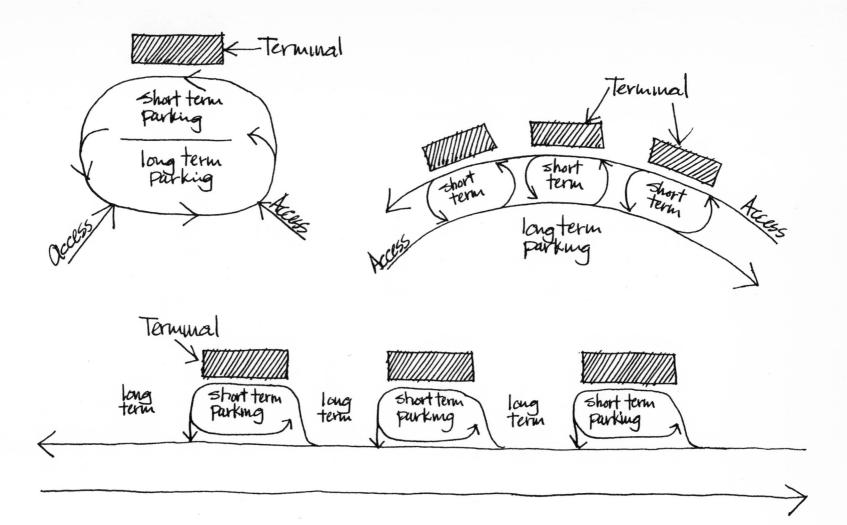

USER ACTIVITIES

Certain building design problems sometimes deserve research. But finding the time and resources for research is difficult. I have found that the application of networks, as at the left, often produces quick insights into a design prob-

lem. In investigating prototypes for basic airport layouts, an examination of the sequence of user activities reveals a large number of prebuilding entry decisions and alternatives. This might suggest the need for prototypes based on site rather than building configuration.

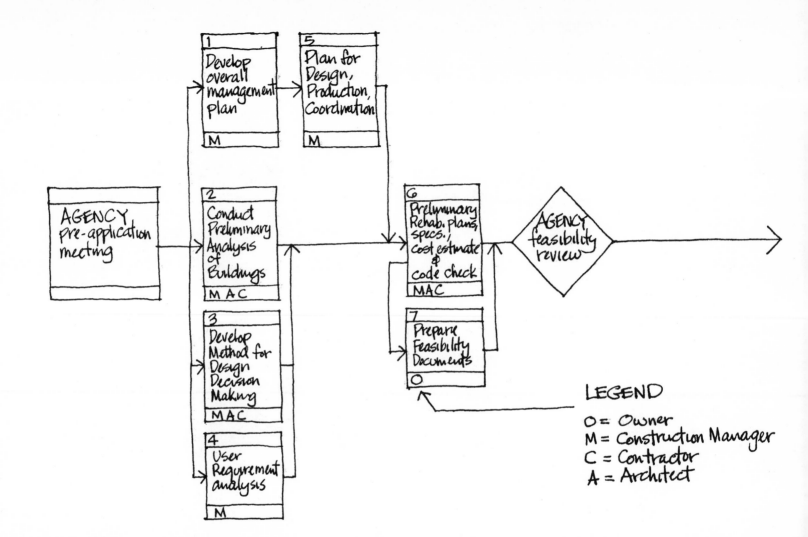

Box 1
1
Develop overall management plan
M

Box 5
5
Plan for Design, Production, Coordination
M

Agency box
AGENCY Pre-application meeting

Box 2
2
Conduct Preliminary Analysis of Buildings
M A C

Box 3
3
Develop Method for Design Decision Making
M A C

Box 4
4
User Requirement analysis
M

Box 6
6
Preliminary Rehab. plans, specs., cost estimate & code check
M A C

Box 7
7
Prepare Feasibility Documents
O

Diamond
AGENCY feasibility review

LEGEND

O = Owner
M = Construction Manager
C = Contractor
A = Architect

HOUSING DEVELOPMENT TEAM

One approach to increasing quality and reducing time and costs in projects such as rehab housing, the example above, is the design or development team, which consists of the principal actors in the building process. The owner, contractor, architect and manager work as a team from conception to realization. A necessary ingredient for success in

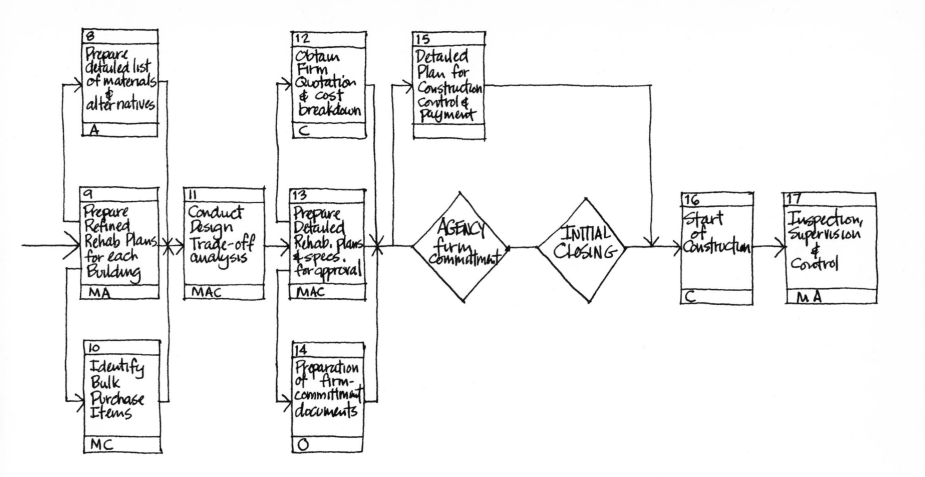

8		12		15	
Prepare detailed list of materials & alternatives		Obtain Firm Quotation & cost breakdown		Detailed Plan for Construction Control & Payment	
A		C		C	

9		11		13		AGENCY firm committment	INITIAL CLOSING	16		17	
Prepare Refined Rehab Plans for each Building		Conduct Design Trade-off analysis		Prepare Detailed Rehab. Plans & specs. for approval				Start of Construction		Inspection, Supervision & Control	
MA		MAC		MAC				C		M A	

10		14	
Identify Bulk Purchase Items		Preparation of firm-committment documents	
MC		O	

what is a new venture for many of us is the under-standing of each team member's role in the process. A relatively simple network can help us to under-stand roles and objectives.

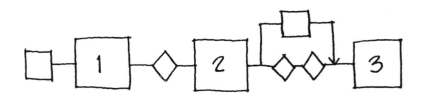

	7	8	9	10	11	12	1	2	3	4	5	6	7	8
LIVING ROOM													■	
FAMILY ROOM				■						■		■		
KITCHEN		■				■					■		■	
LAUNDRY			■											
BATH														
BEDROOM		■	■											
PATIO								■						
TRIPS					■							■		

TIME LOG CHARTS

Many times a simple graphic model of a time log can provide sufficient sense of a situation to allow us to think about it in a comprehensive way. On the left is shown the use of spaces by a housewife on a typical day. The vacant areas are sometimes as informative as those filled in. On the opposite page are examples taken from construction management. The first is a time study of two pieces of equipment and the operator of the loading shovel. The second example shows a means of matching construction foremen with jobs both present and planned.

OPERATOR		TRUCK		LOADER	
Run machine				load truck	
Idle		truck change		Idle	
Run Machine		Idle		load truck	
		truck change		move ahead in cut	
				load truck	
Checking Grade					
Run machine					

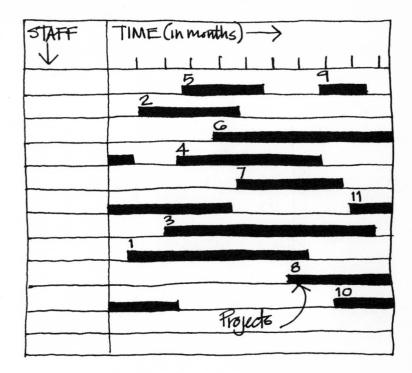

STAFF ↓   TIME (in months) →

Projects

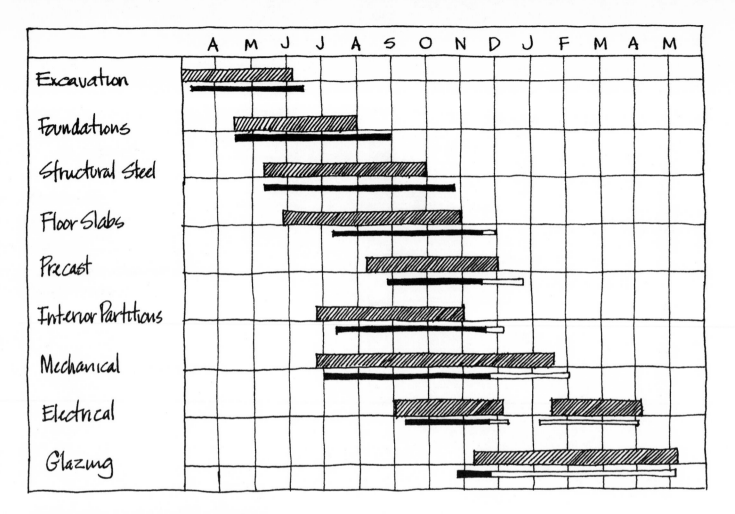

	A	M	J	J	A	S	O	N	D	J	F	M	A	M
Excavation														
Foundations														
Structural Steel														
Floor Slabs														
Precast														
Interior Partitions														
Mechanical														
Electrical														
Glazing														

CONSTRUCTION PROGRESS

Many contractors are still more comfortable with
bar charts because of their simplicity. The chart
above shows a way to incorporate some of the
progress checks that are built into the network
diagrams. The wide bar indicates the job
schedule as planned. The thin bar shows the

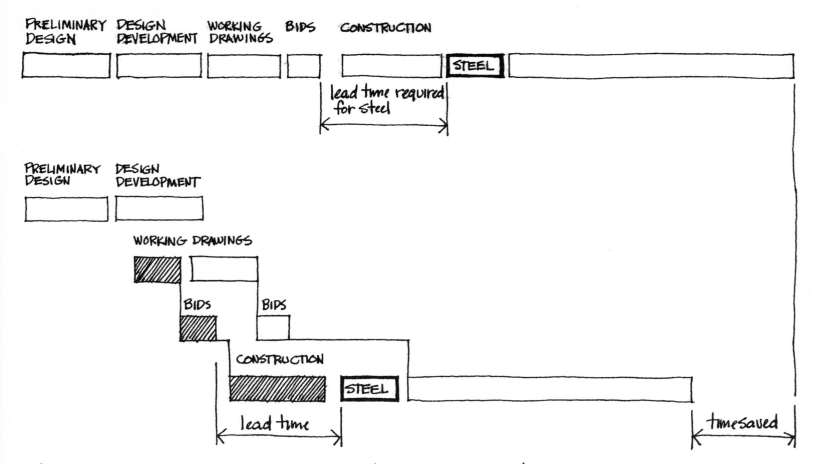

PRELIMINARY DESIGN | DESIGN DEVELOPMENT | WORKING DRAWINGS | BIDS | CONSTRUCTION

STEEL

lead time required for steel

PRELIMINARY DESIGN | DESIGN DEVELOPMENT

WORKING DRAWINGS

BIDS | BIDS

CONSTRUCTION

STEEL

lead time

time saved

actual schedule under which tasks were completed. The blank areas of these bars are projections of task completion based on current progress.

PRE-BIDDING

This is one form of fast-tracking, an approach to reducing design and construction time. In this case a rehabilitation project calls for the installation of a pre-fabricated steel structure soon after the start of construction since foundation work and some of the usual site preparations are not required. Pre-bidding not only permits the shortening of on-site time but also can reduce the total project time.

Specification

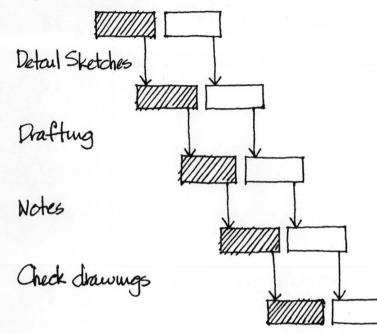

Detail Sketches

Drafting

Notes

Check drawings

FAST TRACK
The principle of fast-tracking can also be applied at a simpler or smaller scale. By dividing the drawings into packages a significant time savings might be realized. The use of a diagram such as the one above might promote the re-examination of the way design development decisions are made and how that affects the use of valuable staff time.

Decision Trees

This particular form of network was most extensively developed in the area of marketing research, where it was applied to both consumer behavior and management planning.

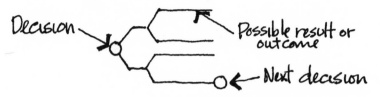

Decision → Possible result or outcome → Next decision

A decision tree consists of two parts: decisions and possible outcomes. Because there are usually two or more possible outcomes for each decision the diagram tends to branch out like a tree. The purpose of this type of diagram is to get an advance look at the possible end results of a series of decisions. When probabilities are applied, it is possible also to estimate the probability of success given two or more initial decisions. The decision tree tool can be used in the area of classic decision analysis or in a number of other ways such as are indicated in this section.

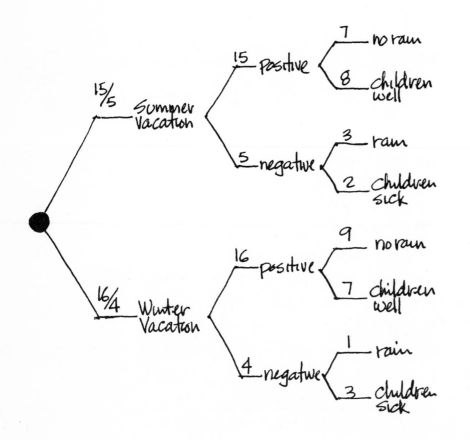

BASIC DECISION ANALYSIS

The diagram at left is concerned with deciding between a summer or winter vacation. Based on the factors of childrens' health and weather conditions the odds for a successful vacation are good for both seasons but just slightly better in the winter. (This might suggest that there are more important factors to consider.)

On the right we look a little farther to see the effects of the second level of decisions. Again by applying probability a weighting can be assigned to the choices in the initial decision. We can also compare parallel routes of circumstances.

146

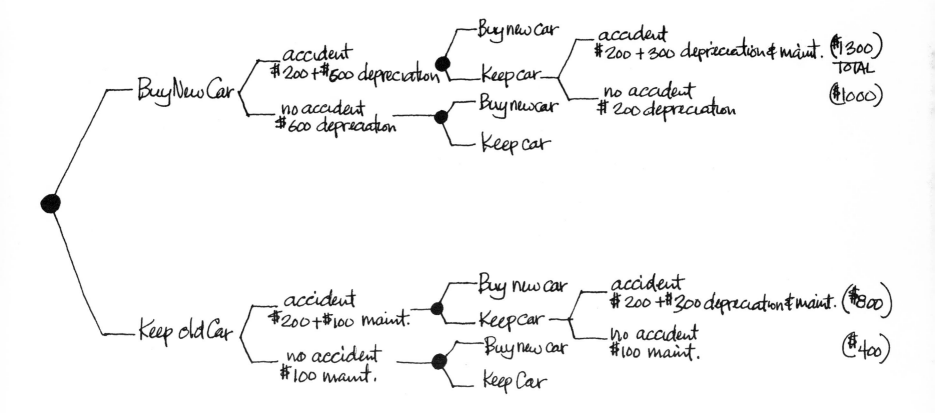

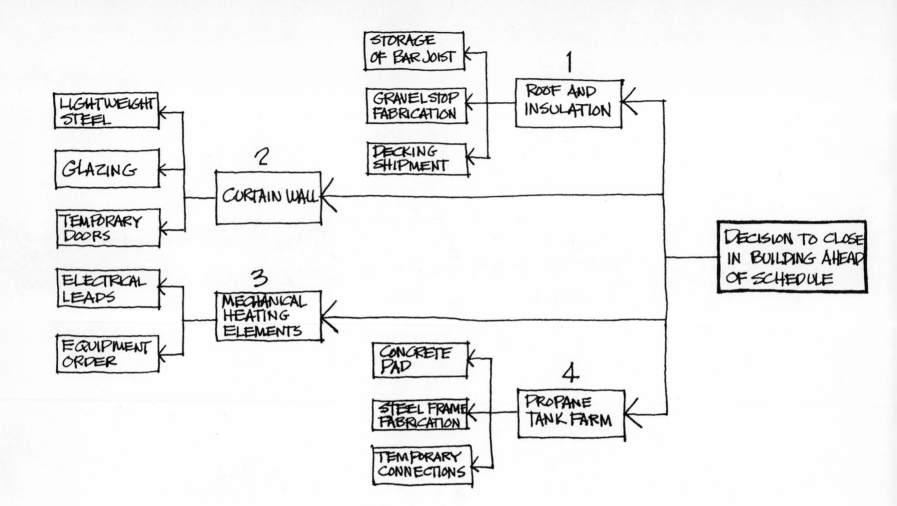

STORAGE
OF BAR JOIST

GRAVEL STOP
FABRICATION

DECKING
SHIPMENT

1
ROOF AND
INSULATION

LIGHTWEIGHT
STEEL

GLAZING

TEMPORARY
DOORS

2
CURTAIN WALL

DECISION TO CLOSE
IN BUILDING AHEAD
OF SCHEDULE

ELECTRICAL
LEADS

EQUIPMENT
ORDER

3
MECHANICAL
HEATING
ELEMENTS

CONCRETE
PAD

STEEL FRAME
FABRICATION

TEMPORARY
CONNECTIONS

4
PROPANE
TANK FARM

IMPACT OF RESCHEDULING

It is nice when all things go as planned on a construction job but change, slow-up and speed-up seem to be the norm. There are just too many things over which the builder has no control. In a hypothetical case the builder has construction in progress on a supermarket. A shift in economic conditions leaves insufficient money for extensive winter construction procedures as originally budgeted. A possible solution is the closing in of the building ahead of schedule. By using a decision tree in a reverse order four major preceding tasks can be identified. In turn each task is checked to see what other tasks or events are important to completing the major tasks on time. This then generates a check list of adjustments to be made in the construction operations.

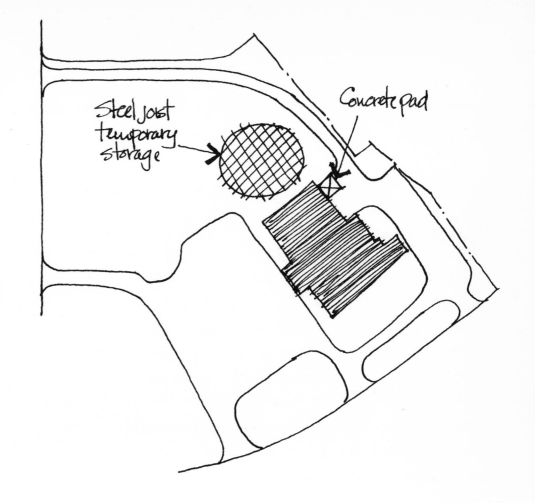

Steel joist temporary storage

Concrete pad

DISPLAY OF OPTIONS

The decision tree diagram assists by providing a framework upon which we can display several choices and at the same time categorize those choices. The first example is a quick analysis of a housing development solution which will help the developer understand consumer choices being offered. This can later be used as reference for potential housing customers. The second diagram relates resources to objectives, namely building configurations. In the third example, at the far right, the purpose is to show the number of approaches or combinations of methods available to meet the need for additional space. Each combination, having different opportunities and constraints, is to be tested against time, cost and quality objectives.

AMENITIES

1 Bedroom
- low rise — balcony / skylight
- high rise — balcony / sleeping alcove

2 Bedroom
- low rise — balcony / direct access
- high rise — balcony / exterior exposure / 2 adjacent sides

3 Bedroom
- low rise — balcony / private court
- high rise — 2 balconies / 2 story living room

4 Bedroom
- low rise — private court
- high rise — 2 balconies

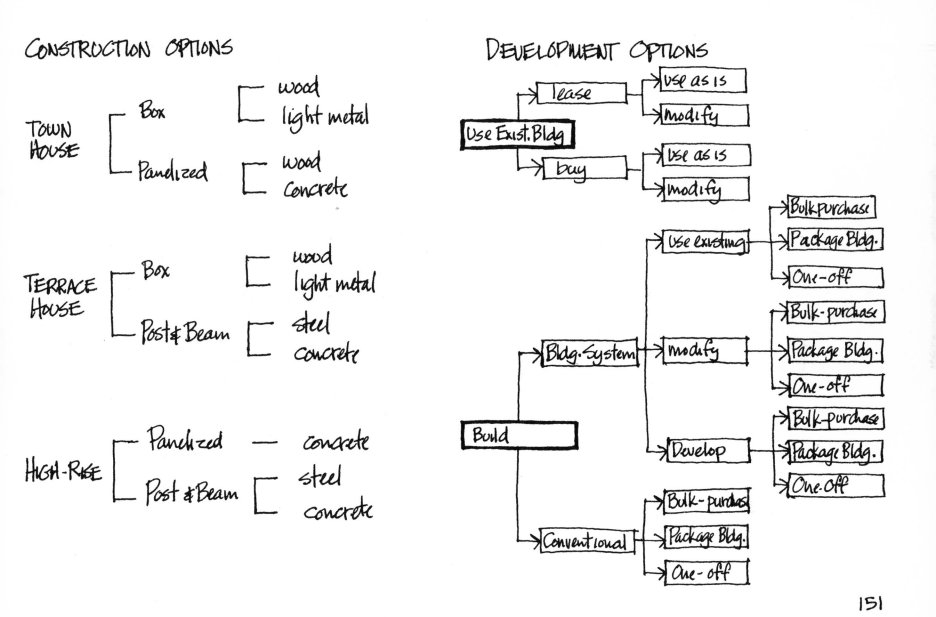

CONSTRUCTION OPTIONS

TOWN HOUSE
- Box
  - wood
  - light metal
- Panelized
  - wood
  - concrete

TERRACE HOUSE
- Box
  - wood
  - light metal
- Post & Beam
  - steel
  - concrete

HIGH-RISE
- Panelized — concrete
- Post & Beam
  - steel
  - concrete

DEVELOPMENT OPTIONS

Use Exist. Bldg
- Lease
  - Use as is
  - Modify
- Buy
  - Use as is
  - Modify

Build
- Bldg. System
  - Use existing
    - Bulk purchase
    - Package Bldg.
    - One-off
  - Modify
    - Bulk-purchase
    - Package Bldg.
    - One-off
  - Develop
    - Bulk-purchase
    - Package Bldg.
    - One-off
- Conventional
  - Bulk-purchase
  - Package Bldg.
  - One-off

151

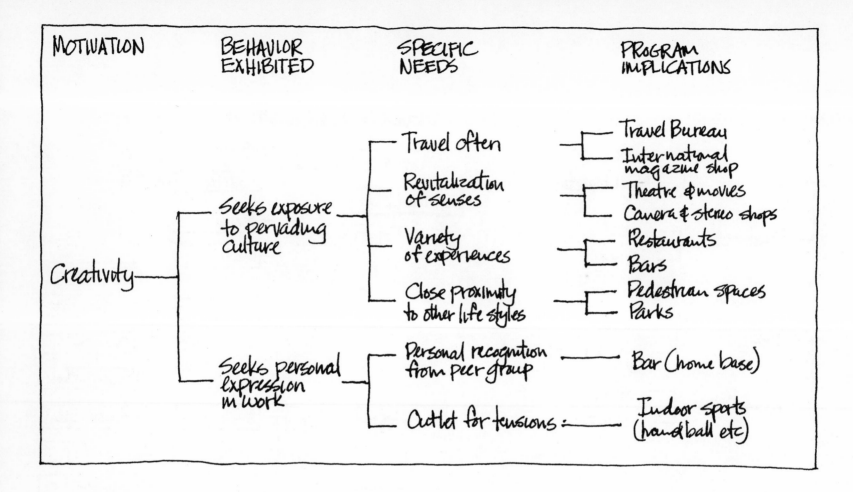

MOTIVATION	BEHAVIOR EXHIBITED	SPECIFIC NEEDS	PROGRAM IMPLICATIONS
Creativity	Seeks exposure to pervading culture	Travel often	Travel Bureau
			International magazine shop
		Revitalization of senses	Theatre & movies
			Camera & stereo shops
		Variety of experiences	Restaurants
			Bars
		Close Proximity to other life styles	Pedestrian spaces
			Parks
	Seeks personal expression in work	Personal recognition from peer group	Bar (home base)
		Outlet for tensions	Indoor sports (handball etc)

DESIGN REQUIREMENTS

One specific contribution that research offers us as developers or designers is simple checklists of design considerations. Our capacity to respond depends upon resources available but at least a checklist of design requirements gives us a goal. The example above is based upon a psychological concept of a hierarchy of related factors starting with a basic motivation and ending with specific implications for community design.

Conclusions

As with any book of this type the ending point is arbitrary and the conclusions temporary. I have set up sort of filing drawers into which I have thrown some examples; there is plenty of room left in these drawers. This is not so much an argument for another book as it is a plea that the reader or "looker" take a certain attitude toward this book. To cite a parallel situation: When I attend a lecture the notes I take are as concerned with the ideas that are stimulated in me as they are with the lecturer's ideas. A lecture or a book can be seen as the beginning of a conversation that each of us will continue with ourselves.

The shortcomings of this book probably lie in the lack of clarity or communicative ability of some of the examples and in failure to reach a sort of silent objective to the degree I had hoped, namely opening further channels of communication between developers, builders, manufacturers and architects.

What the book demonstrates, I feel, is: 1) that a wide range of problems can be "looked" at; 2) that looking at these problems tells us something we did not understand before; 3) that we don't have to accept an all-or-nothing approach to the kind of techniques in this book; we can make them fit our problems.

Well what do we do with this book? Here are some specific recommendations:

1. Do leave the book around where you might stumble over it from time to time.
2. Do use the index to help you discover diagram types with which you are unfamiliar.
3. Do stick pieces of paper in the book when you come across an idea that is particularly useful.
4. Do make notes, especially visual ones, if this book stimulates some ideas.
5. Do try out some of these diagrams on something in which you are presently involved.
6. Don't make diagrams any bigger than those you find in this book. It is important to be able to see the whole diagram at once at normal reading distance.
7. Don't get formal about it. We all know we can do a good presentation of just about anything, but these are sketches. Clarity and speed are what count.

8. Don't be discouraged if the first sketches fail to yield immediate results. Visual thinking is a process of seeing as well as drawing and it may take time to develop perceptions about these kinds of problems.

It certainly is news to no one that man faces a series of crises of which energy, food supply and population are the most immediate — crises which will ultimately test man's ability to marshal his unique talents to avoid extinction as a species. Whether these crises are of global or community scale, it is clear that many of them are the result of our failure to see the interrelatedness of many problems. Solutions and problems are also related. The solution of one problem may often be the cause of a second problem, and, as in the case of pollution, there is simply no room left to dump the unwanted side effects of our decisions. Ten years of work on the urban crisis in this country makes it evident that we cannot just apply or modify old solutions. These problems are not just extensions of old problems but are fundamentally different.

Those responsible for our man-made environment, particularly architects, have a unique responsibility now and in the near future: first because the built environment has already demonstrated its potential for disastrous effect upon our natural environment, and second because of the special set of skills developed through architectural education and practice that can be brought to bear on these new problems. What are these skills and how do they relate to graphic thinking?

Problems must first be understood and then new solutions must be formed. This calls for perception and creativity. In his book The Nature of Design David Pye clarifies perception and creativity and their critical relationship to each other. "Invention can only be done deliberately if the inventor can discern similarities between the particular result which he is envisaging and some other result which he has seen and stored in his memory.... An inventor's power to invent depends on his ability to see analogies between results..." (The emphasis by underlining has been added by me.)

Although general systems theory represents the fundamental shift in perception required, the continuous development of new perceptions of problems is necessary to creativity. Graphics plays a very important part in this process if we can open up to its possibilities. If the images before us can become an integral part of our thinking, perceptions can be changed rapidly while retaining a grasp of the total problem. This is the critical contribution of architecture. We can meet our responsibilities as professionals if we are willing to broaden or shift the boundaries of our traditional concerns.

References

Adams, James L. Conceptual Blockbusting.
    San Francisco: W.H. Freeman and Company, 1974.

Arnheim, Rudolf. Art and Visual Perception.
    Berkeley: University of California Press, 1971.

Ball, John and Byrnes, Francis C. Research Principles
    and Practices in Visual Communication. Wash-
    ington, D.C.: National Education Association, 1960.

Broadbent, G. and Ward, A., eds. Design Methods in
    Architecture. New York: George Wittenborn, Inc., 1969.

Bursk, Edward C. and Chapman, John F., eds. New
    Decision-Making Tools for Managers. Cambridge,
    Mass.: Harvard University Press, 1963.

Churchman, C. West. The Systems Approach. New
    York: Delacorte Press, 1968.

Clough, Richard. Construction Project Management
    New York: John Wiley & Sons, Inc., 1972.

Drucker, Peter F. Concept of the Corporation. Rev. ed.
    New York: John Day Company, 1972.

Kotler, Philip. Marketing Management. Englewood
    Cliffs, N.J.: Prentice-Hall, Inc., 1967.

Lockwood, Arthur. Diagrams. New York: Watson-
    Guptill, 1969.

McLuhan, Marshall. Understanding Media. New
    York: McGraw-Hill, 1964.

Morgio, Mathew. Communication Graphics. New
    York: Van Nostrand Reinhold Company, 1969

Oxley, R. and Poskitt, J. Management Techniques
    Applied to the Construction Industry. London:
    Crosby Lockwood & Son Ltd., 1968.

Ralph M. Parsons Company. The Apron Terminal
    Complex. Washington, D.C.: Department of
    Transportation, Federal Aviation Administration,
    1973.

Pena, William M. and Focke, John. Problem Seeking.
    Houston: Caudill Rowlett Scott, 1969.

Priluck, Herbert and Hourihan, Peter. Practical C.P.M.
    for Construction. Duxbury, Mass.: R.S. Means
    Co. Inc., 1968.

Pye, David. The Nature of Design. New York:
    Reinhold Publishing Corp., 1964.

Wright, Andrew. Designing for Visual Aids. New York:
    Van Nostrand Reinhold Company, 1970.

Index

About the Author

Paul Laseau is currently Acting Director of the Ohio University School of Architecture. Previously he held posts as a designer with Henri Colboc and Gerard Phillipe, Architectes, Paris, France and as project architect with Marcel Breuer and Associates, Architects, New York City. Mr. Laseau is co-founder of Building Science, Inc., a building process consultation firm. Earlier teaching positions include that of instructor of design at the School of Architecture and Environmental Design, State University of New York at Buffalo, and assistant professor of theory and design at the Ohio University School of Architecture.

Holder of a Bachelor of Architecture degree from The Catholic University of America and a Master of Architecture degree in systems building design from the State University of New York at Buffalo, Mr. Laseau is a Registered Architect in the state of New York.